"This book is educational, hilarious, and it makes me wish it was legal to participate in cage fighting with an endangered species."

—MATTHEW INMAN,
CREATOR OF THEOATMEAL.COM AND
AUTHOR OF *5 VERY GOOD REASONS TO PUNCH A DOLPHIN IN THE MOUTH*

"A veritable cavalcade of badly behaved beasts. I couldn't keep my paws off it!"

—DANNY BECK,
AKA SIR PILKINGTON SMYTHE ESQ., EVERSOSTRANGE.COM

"Lombardi may be the anti-cute-and-fuzzy. Her book rightfully reminds us of animals' true, often cantankerous, horny, and self-serving natures."

—JENNIFER HOLLAND,
AUTHOR OF *UNLIKELY FRIENDSHIPS*

ANIMALS BEHAVING BADLY

Boozing Bees, Cheating Chimps,
Dogs with Guns, and Other Beastly True Tales

LINDA LOMBARDI

A PERIGEE BOOK

A PERIGEE BOOK
Published by the Penguin Group
Penguin Group (USA) Inc.
375 Hudson Street, New York, New York 10014, USA
Penguin Group (Canada), 90 Eglinton Avenue East, Suite 700, Toronto, Ontario M4P 2Y3, Canada
(a division of Pearson Penguin Canada Inc.)
Penguin Books Ltd., 80 Strand, London WC2R 0RL, England
Penguin Group Ireland, 25 St. Stephen's Green, Dublin 2, Ireland (a division of Penguin Books Ltd.)
Penguin Group (Australia), 250 Camberwell Road, Camberwell, Victoria 3124, Australia
(a division of Pearson Australia Group Pty. Ltd.)
Penguin Books India Pvt. Ltd., 11 Community Centre, Panchsheel Park, New Delhi—110 017, India
Penguin Group (NZ), 67 Apollo Drive, Rosedale, Auckland 0632, New Zealand
(a division of Pearson New Zealand Ltd.)
Penguin Books (South Africa) (Pty.) Ltd., 24 Sturdee Avenue, Rosebank, Johannesburg 2196, South Africa
Penguin Books Ltd., Registered Offices: 80 Strand, London WC2R 0RL, England

First edition: October 2011

Library of Congress Cataloging-in-Publication Data

Lombardi, Linda, 1961–
Animals behaving badly : boozing bees, cheating chimps, dogs with guns, and other beastly true tales / Linda Lombardi.
p. cm.
"A Perigee book."
Includes bibliographical references and index.
ISBN 978-0-399-53697-7
1. Animal behavior—Anecdotes. 2. Animal attacks—Anecdotes. I. Title.
QL751.L64 2011
591.5—dc23 2011018703

PRINTED IN THE UNITED STATES OF AMERICA

10 9 8 7 6 5 4 3 2 1

CONTENTS

INTRODUCTION: Not As Cute As They Want You to Think — vii

1. Muggers, Burglars, and Thieves — 1
2. Assault, Running Amok, and Arson — 13
3. Kinky Creatures — 29
4. Animal Family Values — 43
5. Party Animals — 55
6. Beastly Devices and Deceits — 73
7. Masters of Misdirection — 91
8. With Friends Like These . . . — 107
9. Our Own Worst Enemies — 123
10. Ungrateful Beasts — 137
11. Heroic Humans — 155

Acknowledgments 171

References 173

Index 201

INTRODUCTION

Not As Cute As They Want You to Think

ANIMALS. WHY DO WE THINK THEY'RE SO CUTE, SO NOBLE, SO admirable? They eat poop, for Pete's sake. Some of them will bite your head off as soon as they look at you. Yet somehow we believe that animals are actually better than us. How did they figure out how to get such good press?

Sure, you can try to make excuses for some of these creatures. Maybe the bear that mugged a New Jersey man for his Italian sandwich in his own driveway was starving. Maybe the turkeys who were attacking people in a Philadelphia suburb—at the end of November—had a point. Maybe all those "innocent" wild things are defending an increasingly precarious existence. But dig a little deeper, and you'll discover behavior that seems less excusable as well as oddly familiar:

- Bees love alcohol, even, says one researcher, more than college students do.

- A rabbit who lives in a pub in England is addicted to gambling with a slot machine.
- African elephants raised by teenage mothers form violent youth gangs.

There's a lot that animals don't want you to know. In fact, the better their public image, the darker their secrets: lazy, infanticidal lions; hummingbird rapists; and of course, our own dogs, who eat our money, set our houses on fire, and—in more than one case—actually shoot their owners with guns. Think of almost any human vice or crime, and you'll find that animals do it too. Let's look at a few examples:

- **Stealing:** Theft is one animal crime that can sometimes seem justifiable. At least when a bunch of elephants in Thailand gang up to stop a truck full of sugarcane and rob it, what they're stealing is edible. But how do you make excuses for a fox—already born with a fur coat—who stole a sweater on a ninety-degree August day in Virginia?
- **Assault:** While some animal attacks seem like plain hooliganism, others feel personal, perhaps even editorial. You almost can't blame the groundhog who bit the mayor of New York City on a Groundhog Day just after he'd made massive cuts in funding for the city's zoos. But even if it were a political statement, isn't the cow who toppled the blind, elderly British MP—on the man's birthday, no less—going a little too far?
- **Perversion:** Humans assume we're the sole species that has sex purely for enjoyment. But if animals mate only to pass on

their genes, some of them sure are doing it wrong. Some male frogs pile up on a female till she drowns, and a rooster in England was in the news when his mate died of exhaustion from his enthusiasm—not great approaches if the point really is to produce offspring. We'll also see in Chapter 3 that it really is true that *everyone* masturbates.

- **Infidelity and domestic violence:** If you've ever felt bad about your own dysfunctional family, don't—you've got plenty of company. People may have trouble with monogamy, but they probably do better than the bird species where 95 percent of the paired-up females commit avian adultery. And it's not just birds; all over the animal kingdom we find lousy parents. Human dads may be reluctant to help with the kids, but at least they don't kill a female's babies so they can sire their own young. The much-vaunted maternal instinct isn't all it's cracked up to be either. Some animals abandon their young as a matter of course, but don't feel too bad for the many babies who are left to fend for themselves; at least their mothers didn't eat them.

- **Substance abuse:** You might assume that drinking and drug use require a certain advanced level of civilization, but humans didn't invent fermentation. Alcohol occurs in nature, and drunk animals behave just as badly as their human counterparts; in fact, their reactions are often quite familiar. Bees, for instance, have been observed throughout history getting drunk on fermented nectar. They stop doing their jobs, stumble around, have flying accidents, pass out, and may even be ejected from the hive by disgusted relatives.

  Animals also use many other natural intoxicants. Once

you read Chapter 3, you'll never look at that apparently innocent catnip mouse in the same way again. And if you're hooked on candy and chips, you've got creature company: Neuroscientists have discovered that rats can develop a Ho Hos habit.

- **Lying:** Okay, so animals steal, do drugs, are perverts, and have lousy family lives. But surely there are still some kinds of bad behavior that are sophisticated enough to separate us from the beasts? Surely, only *we* have the smarts to lie, right?

  In reality, nature came up with deception long before the first human told his mate that she looked great in that goatskin dress. Some animals trick other species into raising their young, and others literally cry wolf to fool their buddies into running away from some tasty food. Of course, when it comes to lies, smarts don't hurt. As you'll see in Chapter 6, our closest relatives can plan ahead and even construct tools to help them lie, like the orangutans in Borneo that use leaves, held up to their mouths, to make their calls sound deeper. The low tones fool predators into thinking the apes are much bigger than they really are because—like on an Internet dating site—interactions in the rain forest often aren't face to face.

ANIMAL IGNORANCE IS NOT BLISS

The mere existence of bad behavior in animals is just the start. It would be one thing if people understood what we're up against. But the real problem? We humans are completely deluded about the true nature of our fellow creatures.

Take dolphins for instance. What animal is more mystical and profound than the dolphin, swimming through the ocean surrounded by serene new-age music and wind chimes? The ugly truth behind that charming smile, however, is that they are gang rapists who kill babies for fun (see Chapter 7).

And we sure don't know how to choose a best friend. The misdeeds of canines are extensive enough to get a whole chapter to themselves (see Chapter 8).

Now, you might ask, what's the harm in having beautiful illusions about our fellow creatures? Maybe we shouldn't throw stones. Maybe every species deserves its own secrets about whether it really, say, mates for life. And who gets hurt if you think that those big brown puppy dog eyes mean "I love you" rather than "Give me a cookie, *now*"? The dog gets a cookie either way, right?

The problem is, while we have the wildest misconceptions about them, animals have got our number. In the *New York Times*, primate researcher Frans B. M. de Waal said that chimps and orangutans in captivity lure humans to come close by holding out a piece of straw and making a friendly face:

> "People think, Oh, he likes me, and they approach," Dr. de Waal said. "And before you know it, the ape has grabbed their ankle and is closing in for the bite. It's a very dangerous situation."

The conclusion is inescapable: Believing that animals are charming, adorable, noble, and true is not a victimless crime. If you don't want to be part of the problem, read on and learn the truth.

ONE

Muggers, Burglars, and Thieves

AS CRIMES GO, STEALING IS PRETTY MUCH UNSKILLED LABOR. You don't have to be some kind of evil mastermind to snatch someone else's stuff: All it takes is opportunism and a blatant disregard for others. If those are the qualifications for the job, animals are the perfect candidates, so it's no surprise that they commit this entry-level offense frequently.

Now, at first glance, it can be easy to make excuses. Animals need to eat. And since they don't generally have money, what's a poor hungry creature to do? You'd have to be downright mean to begrudge a seagull that French fry you dropped on the boardwalk. And what do campers expect when they leave tempting food sitting around unguarded? Of course the bears are going to want a share.

But not all animal theft, even of food, is so easy to excuse. Sure, it's one thing if we're vacationing on the bear's home turf. But how about the man who was mugged on his own property for his Italian sandwich? Henry Rouwendal of Vernon, New Jersey, was in his driveway when it happened:

> Suddenly, Rouwendal said, something hit him from behind.
>
> "It blind-sided me. I was on the ground and I was thinking, 'What the hell just hit me?' " said Rouwendal, who also suffered a large cut on his left temple and several deep bruises on his knee, elbow and buttocks.
>
> Rouwendal [was] knocked, face-first to the ground. When he rolled over, the bear was standing over him and then grabbed the sandwich.

It's hard to believe that starvation is the motivation when some of these animals are so fussy. In this case, the bear took the bread and cold cuts but left the lettuce, onion, and tomatoes behind.

There was also the seagull in Aberdeen, Scotland, who became something of a celebrity for his shoplifting habits. Far from being satisfied with whatever might fill its belly, the bird would stroll into a store and take a bag of chips—but only cheese Doritos, its favorite kind.

You might argue that a bird can't understand it's committing a crime. An animal doesn't know that chips on a store shelf are different from the fries and ice cream dropped on the sidewalk, right? On the contrary, the shopkeeper says that the gull is careful to make sure there are no witnesses:

> He's got it down to a fine art. He waits until there are no customers around and I'm standing behind the till, then he raids the place.

UNINVITED COMPANY

Along with mugging and daylight robbery, breaking and entering is another foraging technique you won't usually see on the natural history documentaries. One man in Montana was woken in the small hours of the morning by a commotion and discovered that a bear had opened the porch door and was raiding his chest freezer:

> He was about four feet from me with his head in the freezer munching on a pizza. . . . It was almost comical if it wouldn't have been that close.

This man should probably be thankful, though, because a sneaky midnight raid is better than some of the alternatives. In the upscale suburbs of Cape Town, South Africa, baboons have taken to committing outright home invasions. A 2006 report quotes anti-primate crusader Joan Laing, cochair of the Welcome Glen Baboon-Free Neighbourhood Action Group:

> "They break windows to get into houses," Laing said. "They even know how to open doors. And once inside, they make a mess. They empty the fridge, ruin furniture, and defecate all over."

These raids take place in broad daylight, and the monkeys don't wait for the house to be unoccupied. Laing has suffered these invasions several times when she was at home. "They simply brushed past me. I had to get out of the way," she said.

The result can be damage to more than property. A twelve-year-old boy was confronted by a troop that broke into his home in the summer of 2010:

> Hearing noises from the kitchen, he went to investigate and found the beasts ransacking cupboards. When the child fled upstairs to find his babysitter, three males gave chase and surrounded him as he made a tearful phone call to his mother, while the animals pelted him with fruit.

"When he called me he was terrified. They had him surrounded," said the traumatized child's mother. Sadly, apparently the existence of anti-baboon activist organizations has done little to solve the problem, and no doubt children will continue to suffer such lifelong psychological damage, as civilized life in the neighborhood has almost become a thing of the past:

> "Lunch parties in the garden are now just impossible," a homeowner complained. "It is so unrelaxing. Rather than chatting over our meal, we are looking over our shoulders and bolting the food as quickly as we can before it is stolen. We can't even leave a window open in summer. We are under siege."

And animals making this type of brazen daytime raid have no more sympathy for other animals than they do for humans. In New Hampshire, a bear invaded a home for a meal of fruit and Pepperidge Farm crackers, but the latter weren't the only goldfish that suffered. The homeowner arrived just in time to save the family pets, one of which was left flopping around on the counter when the bear drank from the fishbowl. (The thief had second thoughts about abducting what he may have believed was a relative—the stuffed teddy he took from the house was left behind in the yard.)

TINY TERRORS

Don't kid yourself into thinking you're safe from home invasions because your neighborhood is free of big scary wildlife like bears and baboons. In Britain, some chipmunks escaped from a wildlife park in 2005 to devote themselves to a life of crime. Their campaign of terror was chronicled in detail by the *Sun* newspaper, which provided chipmunk recipes for citizens wanting to take justice into their own hands. In one case, a family in Kent found one of the rodents in their kitchen, brazenly eating their breakfast cereal:

> It was terrifying. I've never seen anything like it. We could hear it banging about in the kitchen and when my husband went in it had its head in the cereal and was munching away. . . . My cats are traumatised. We're too scared to let them out in case it's still lurking.

If you think this is an overreaction to finding a small rodent in your kitchen, read how another woman was attacked in her own backyard:

> It was staring right at me and wasn't frightened of me at all. The next thing I knew it was jumping straight towards me

> and went on the attack. . . . I thought it was going to bite me. I screamed my head off and ran for the kitchen door, then banged it shut and sat quaking inside.

These may seem like the exaggerations of people who are unfamiliar with the nonnative wildlife in question, or your suspicions may be aroused by the fact that the *Sun* states on their website that they pay money for stories. But the danger was confirmed by a professional naturalist who warned about the chipmunks: "They will stop at nothing to get what they want."

Other small animals that seem friendly and charming may deliberately soften people up before victimizing them. A magpie in a German village started out by making friendly overtures, perching on people's shoulders and being fed nuts, before the situation took a darker turn. First, the bird made friends with the local postwoman only to steal her ballpoint pen; then, the situation escalated, as one resident described:

> She decided she wasn't getting enough and just started taking what she wanted. Now she flies straight into my living room through the window, eats my marzipan potatoes and chocolates on the table and sits on the sofa.

GRAND THEFT ANIMAL

Our vehicles aren't safe from animals either: both bears and baboons, for example, seem to be increasingly targeting cars as well as houses. In 2009, deputies in Colorado responded to reports of a car theft in progress to find a bear in the driver's

PUP AND PRIMATE PREMEDITATION

Animals that steal food aren't necessarily overwhelmed by hunger and unable to overcome an impulse to grab—sometimes there's quite a bit more planning involved.

Staff at an animal shelter in Britain came to work one morning to find a canine party in progress: The kitchen had been raided of treats and toys and the dogs were running loose, the doors of their kennels somehow unlocked. The even bigger surprise was that it happened again and again, with more dogs joining the fun each time. When the caretakers set up surveillance cameras, they found that a dog called Red had figured out how to put his teeth through the bars of his kennel to release the spring-loaded catch on the door—and then, once out, he'd do the same for his friends and they'd raid the cupboards and spend the night carousing.

And those South African baboons who break into cars don't waste time trying all the doors to see which are unlocked—they've come up with a more efficient approach. They listen for the beeping of remote door locks. Says a Cape Town city official, "They're waiting for the sound of the car alarm. If they don't hear the 'tweet tweet' they make for the door."

seat, rummaging around in the passenger compartment. To make matters worse, their reaction was to open the car door to help the bear escape—after making sure to take photographs first. So the bear gets the peanut butter sandwich left in the car and Internet fame, and how does it thank the owner?

> It left a foul-smelling "present" on the front seat.
>
> The Toyota was trashed, with its air bags, seats and stereo torn to shreds. It's a total loss.

Ominously, bears seem to be making progress in their understanding of how cars work. In 2010, when officers arrived to investigate a car honking and making a commotion, it turned out to be a bear who had managed this time to drive the vehicle over a hundred feet from its original parking place.

Incidents of this type will no doubt continue, since the police response was again to open the doors and assist the culprit to escape. However, even where authorities have taken a harder line, there's been little progress, despite perpetrators who are well known by name. In South Africa, gangs of baboon have learned to open doors to steal food and valuables from parked cars. A particular gang leader by the name of Fred is an expert, as described by an official tasked with tracking the troop:

> He'll hit four or five cars in like five minutes. Fred's operation is to open car doors. He leaves normally with a handbag. Until he's satisfied he's got all the food, don't try to get the bag back.

The baboons will attack if confronted; Fred has bitten people—and he's a sexist pig as well:

> Shouts and whistles are used, but in tougher baboonhoods like Simon's Town, whips are cracked and crackers set off to move apes back. There are no female monitors as the baboons have proven not to listen to women.

FOOTWEAR FETISH

Food is a necessity, but that's hardly the case with many other things that animals steal. Sure, maybe the monkey who stole five pairs of eyeglasses from an office in India needed them. It's probably hard for a monkey to get an appointment for an eye exam, so he had to try several pairs to get one that was the right prescription. And in some other cases, if the perps were planning to resell the items, the theft might make some kind of sense. But if that were the case, the porcine and canine diamond thieves in Yorkshire, Maryland, probably shouldn't have swallowed the goods. Likewise, dogs steal money fairly regularly, and though you'd think it would be better to exchange it for some tasty steak, they always seem to eat the cash itself, like a dog in North Carolina who gobbled up a whole $400.

In fact, animals steal lots of stuff that they have no practical use for. A fox already has a fur coat, so what's its excuse for assaulting two people to steal a sweater on a ninety-degree August day in Charlottesville, Virginia? Even more inexplicable is the fox in Germany who took over a hundred pairs of shoes, including "muddy hiking shoes, wet Wellingtons, steel-capped workman's boots, flipflops and old slippers," from doorsteps and front porches in the middle of the night. And when her collection was discovered and people took their shoes back, she stole more to replace them.

Elsewhere, cats have stolen whole collections of human garments and accessories. One in Seattle amassed a collection of thirty gardening gloves and his owners put out a bucket where people could come and look for their missing property. The neighbors reportedly thought the situation was funny, but what they didn't realize is that gloves may be just the first step on a slippery slope.

A cat named Oscar in Portswood, Southampton, England, started out with gardening gloves too, but then he moved on to more disturbing items. When reported on in July 2010 he had dozens of socks, several types of gloves, a knee-pad, a paint roller—and a collection of ladies' and children's underwear. Oscar brings home as many as ten items a day, and his owners seem to believe that this disturbing revelation about their losses will be a comfort to Oscar's victims:

> If any readers in the Portswood area are missing the said items of underwear it would be good to put their minds at rest that it's only a cat pinching and not someone more unpleasant.

NO RESPECT

Some might object to making a big deal out of an animal stealing a few old gardening gloves and shoes. Don't be so materialistic, right?

But some of these thefts cause bigger problems. In New Zealand, a Scottish man lost his passport when a kea parrot stole his bag from the luggage compartment of a bus. Stranded far from home without the document, the man had to wait over a month and pay hundreds of dollars for a replacement:

> "Being Scottish, I've got a sense of humour so I did take it with humour but obviously there is one side of me still raging," he said. "My passport is somewhere out there in Fiordland. The kea's probably using it for fraudulent claims or something."

PARTNERS IN CRIME

The impulse to ignore animal crimes is so extreme that human criminals can use it to their advantage. When a plant store in Texas had a series of thefts, they installed a security camera, and found that the culprit who had stolen dozens of plants, flowers, and garden statues was a monkey. But the shop owner declined to press charges because she found the situation "humorous," despite the fact that the monkey was clearly handing off the goods to a person on the other side of the fence.

A common and highly effective animal coconspirator is the pigeon. No one takes much notice of this humble bird, and its ability to fly gives its human accomplices access to otherwise hard-to-reach places. Recently police in Columbia captured a pigeon smuggling marijuana and cocaine into a prison. The police commander commented, "This is a new case of criminal ingenuity," but clearly he just hasn't been paying attention. Two years previously, Brazilian police discovered that pigeons were being used to smuggle cellphones into a jail, and before that, a pigeon was taken into custody after smuggling heroin into a prison in Bosnia.

Not everyone is as inattentive to the potential pigeon menace as the Columbians had been. In 2008, alert Iranian authorities arrested a number of pigeons near a nuclear facility on suspicion of spying. Asked to comment, one diplomat told *Sky News*, "It's clear there has been some sort of coo in Tehran."

And some of these crimes are downright disrespectful. A cemetery in Michigan that decorated the graves of almost a thousand fallen soldiers for Memorial Day was struck by a mystery thief: Many of the flags were taken, leaving just bare sticks behind. The culprit was revealed when a squirrel stole another flag right under the nose of the cemetery's superintendent.

Elsewhere, at a drive-through wildlife park in England, baboons that tired of merely taking car mirrors and wiper blades came up with a new entertainment: They learned how to open rooftop luggage boxes. The contents can't be of much practical use to the monkeys, so no doubt the real value is in the reaction of the victims. As the park manager observes significantly, "Let's face it, nobody wants to see a baboon running up a tree with their underwear."

TWO

Assault, Running Amok, and Arson

WHILE ANIMAL THIEVES HAVE MOSTLY FLOWN AND CRAWLED beneath society's radar, animal violence does a better job of getting our attention. In fact a whole genre of reality TV has sprung up to document the more gruesome creature crimes.

While most of us are at little risk of losing a limb to a crocodile if we follow the most elementary safety precautions, in many cases, the victims of animal attacks are innocently going about their own business. A schoolteacher in an English village was riding her horse along a lane in June 2009 when an attacker came out of nowhere:

> "I heard a loud sound and felt a blow to my head and my helmet," she said. "It was quite a hard blow—I could see stars afterwards."

Other victims of the attacking buzzard that was going after

joggers, cyclists, and dog walkers were not so fortunate to be wearing helmets when they were assaulted.

Innocent recreational activities can be even more dangerous where larger animals are common, and they don't have to be predators. An Australian researcher collected reports of fifteen unprovoked attacks by kangaroos in the space of two years. In one case, a former football player, probably no easy target, was punched unconscious. In another, a kangaroo chased a dog into a pond and tried to drown it, then went after the owner as he tried to rescue his pet. He described his resulting injuries:

> A large gash above my right eye—all that blood left me unable to see. Then there's several large and deep scratches on my face, my neck, my back and chest, which are from the roo trying to push me down into the water.

A study by the Centers for Disease Control and Prevention indicated that about twenty people a year are killed in the United States by cows. These allegedly placid bovines ram, gore, trample, and kick people in the head, as well as crushing them against walls; in one case, a cow administered a fatal dose of antibiotic by knocking down a victim with a syringe in his pocket. A closer look at a few of these incidents uncovered some alarming details:

- In about three quarters of the cases "the animal was deemed to have purposefully struck the victim."
- One of the murderous bulls had been hand-raised and bottle-fed by the victim and his family.

- And finally, watch your back: in at least one case, the victim was attacked from behind.

PICK ON SOMEONE YOUR OWN SIZE

Smallness doesn't stop an animal from going after humans, as demonstrated by the home-invading chipmunks in the last chapter. In Germany, a ten-year-old girl was climbing a tree when a squirrel attacked her. She had to be rescued by firemen when it chased her up into branches almost thirty feet up and she couldn't get down on her own.

Keeping your kids out of potential squirrel territory isn't enough to keep them safe, either: They have gone after children on playgrounds. In Florida, a squirrel attacked a three-year-old on a swing at a daycare center, sending him to the hospital; a state trooper who responded was also treated for injuries. The same squirrel was believed to be responsible for at least seven attacks, with victims that included another child and a man sitting on a park bench. And in San Jose, a squirrel actually ran right inside an elementary school classroom, sending a child and two adults to a hospital for treatment of scratches and bites.

Animals are unashamed to target other vulnerable populations as well. In Melbourne, Australia, a billy goat invaded the grounds of a nursing home. After leaving the sixty-year-old gardener with cuts to the head and arm, it knocked a seventy-year-old man onto the ground and butted him until pulled off by two police constables. And in England, a crow attacking people on a jogging trail confined itself to what it perhaps thought were the easiest marks: The bird only went after blonds.

ANIMAL ARSONISTS

Animals can cause property damage with nothing but the tools nature has given them, and not just the obvious ones. For example, in 2009, cows did $100 worth of damage to a Tennessee home merely by sticking their heads through a fence and licking it, ripping off a screen window, cracking glass, and pulling down a gutter.

Some animals even seemingly commit arson. In Iowa, a goat started a fire that destroyed a home by knocking over a space heater, and in Washington, a rat chewed through the electrical cord of a jukebox at a VFW post and started a blaze that destroyed a collection of antique war memorabilia and caused over $1 million in damage. And sometimes the guilty party poses its own danger as well, as Pittsburgh-area firefighters found when they responded to a fire caused by a space heater:

> *"The first initial report we got was that they found an elevator," North Beaver Township Fire Chief Paul Henry said. "With the breathing apparatus on and the radio traffic that we got, it sounded like 'elevator.' I said, 'No, there can't be one in that building, I know that building,' and they said, 'No, an alligator.'"*
>
> *Henry said that's when he pulled his men out of the burning building.*

However, the animals most likely to start fires are the ones closest to us. Cats on the countertop can do more than annoy, as one family in Washington State found when they were left temporarily homeless by a kitchen fire. Investigators determined that the blaze had originated from a toaster oven, which was found with its switch pressed. The owner said the cat had taken to sleeping on top of the appliance, and—although typically forgiving—had to admit that he "probably did some step aerobics" that had turned it on.

In fact, it's estimated by the American Kennel Club (AKC) that nearly one thousand house fires a year are started by our own pets. Dogs do their share of damage, as well, and the AKC even recommends that dog owners remove or cover stove knobs when they are not in the kitchen, since turning on the stove is the cause of many of these fires.

That advice might have helped the owners of a dog in Wales that jumped up and turned on the switch under a chip cooker, starting a fire that resulted in £6,000 ($9, 700) in damages. Owner Paul Gregson theorizes that the dog was just looking for something to eat:

> *Three-year-old Alfie is a liver-coloured, flat-coated retriever who Paul described as "very lively, bouncy and smelly." He added: "The breed is a cross between a pointer and a red setter. They are slow to mature—if they mature at all. He exists to eat. He's a walking stomach."*

Although owners of any type of dog should be wary of the risks, it's possible that certain breeds may have more of a firebug tendency than others. One owner of a flat-coated retriever was unsurprised by Alfie's story, given that her dog had once turned on the stove and filled the kitchen with gas. "This is why I have childproof knobs on all my burner knobs," she said, echoing the AKC's advice.

NOT SAFE IN OUR BEDS

While journeying into the wilderness may put you at risk of assault by animals, you actually don't even have to leave your own property. One woman in Queensland, Australia, was picking roses in her garden when a kangaroo pushed her to the ground and kicked, scratched, and bit her all over her body. An elderly lady in England was likewise in her yard when a vicious crow took advantage of her helpless position:

> I pulled my sunchair towards the light. As I sat down on it, it tipped up backwards.
>
> While my legs were up in the air this crow came down and started dive-bombing me and making screeching noises.
>
> It was like a horror movie. I got up and started running. I shouted "It's after me," and then I fell in the flower bed.

She escaped with just a sprained ankle, but was left shaken: "Scared would be an understatement," she said. "I had to pour myself a brandy."

Remaining indoors is no safer. We saw in the last chapter that our homes are not safe from animals looking for food, but even more frightening are invaders whose motives seem to be pure mayhem. In 2009, a couple in Australia was woken up by a kangaroo that had come in through a window and was jumping on their bed. By the time the man of the house wrestled the monster out the door, he had "scratched buttocks and shredded underpants," as well as holes gouged in his walls and furniture and two traumatized children.

It's also worth noting that animals are much less forgiving than people of such incursions onto their home turf. The man with the unexpected marsupial in his bed was content to shove the intruder outside and let it hop away. That kangaroo was luckier than the woman who went for a swim at her summer cottage in Wisconsin. She was unruffled by seeing one otter, but when two more appeared, she sensibly "felt uncomfortable and swam to shore." Despite the fact that the woman almost immediately removed herself from their territory, the otters pursued her:

> She said she had her hands on the shore and legs in the water, "and there they were—one on the right leg and one on the left leg."

She was bitten eight or nine times, and had to have rabies shots.

JUST SHOWING UP

Animals can also cause plenty of trouble by simply being in the wrong place at the wrong time.

Imagine combining the vulnerable feeling of having one's pants down with a very common phobia. A man in the Bronx was lucky to notice his visitor, a three-foot-long corn snake, before he sat down on his toilet, as did a woman in Poland—of course, it's hard to miss a six-foot anaconda. But not everyone is so attentive. A man in rural Taiwan made the TV news when he was bitten on his penis, and a woman in Florida was in the hospital for three days after being bitten by the venomous water moccasin in her toilet.

MULTITALENTED MARAUDERS

In Germany, wild boars commit every offense we've seen so far, and then some: blocking roads; rampaging through villages and breaking into shops and homes; and chasing joggers, cyclists, and pedestrians. They cause so many car crashes that a European auto club conducted a crash test with model boars—with spectacular results—to publicize the danger. Nor is public transit immune: In Berlin, some buses avoid stops where boars congregate.

Wild pigs dig up everything from graveyards to gardens to soccer fields. One mob attacked a man in a wheelchair, and a lone boar crashed through the glass door of a church in Frankfurt, right into a meeting of mothers with young children.

They also have no fear of people. Two policemen who responded to a report of a boar in a liquor store gave chase down a nearby street. The police report described what happened next:

> *Initially the attempts of the officers to drive the animal towards the forest appeared to be successful, but suddenly the boar appeared to change its mind, became increasingly aggressive and finally attacked. . . . The two officers were only able to get to safety with a bold leap onto the balcony of an apartment block.*

It's not just the rampaging that makes these pigs a threat: They have smarts as well. Sportsmen say that boars eat only the interior parts of cornfields, leaving the outer edges of the field intact to conceal their activity, and that they've learned how to avoid hunters by hiding when they hear car doors slamming.

Boars also have the advantage of a well-developed social structure, allowing them to benefit from the knowledge of an experienced lead sow, but their society may be following ours down a troubling path, according to one boar biologist:

> *A sow can produce a litter of up to eight piglets per year. . . . And because their reproduction depends on weight rather than age, we're seeing boars of just nine months—mere teenagers—producing young.*

In some cases the boars are an agent of our own technological troubles turned against us. Climate change has increased their numbers, because warmer winters are easier to survive and increase the availability of their favorite foods. And in 2009, almost €425,000 ($555,000) was paid out to German hunters in compensation for wild boar meat that could not be eaten: It was too contaminated by radiation from the Chernobyl disaster that lingers in their diet of mushrooms and truffles.

However dramatic, though, a snake in the bathroom can affect only a limited number of people. The best way to bring whole swathes of society to a standstill at once is to disrupt traffic, especially at rush hour, and here's where animals really show their stuff. In India, elephants sometimes seem to block the road for the pure pleasure of it, like one that brought traffic to a standstill for an hour on a highway in West Bengal in 2010:

> A fully-grown wild elephant, probably from Chapramari Wildlife Sanctuary, had strayed onto the highway and almost lorded over the road not letting any vehicle move an inch ahead.

> The vehicles on the highway had no other option—they were forced to wait until the pachyderm moved away leisurely at its own will.

Some animal roadblocks do more than simply interfere with progress: In Germany, when a woman tried to shoo away five cows that were in her way, one of them climbed onto the car's roof and down the other side, wrecking it. And in the spring of 2010 a Greek highway had to be closed down after a carpet of millions of frogs caused drivers to skid off the road.

RUNNING WILD

Native wildlife is a traffic problem in many places—in Maine, for instance, there are about six hundred collisions with moose per year. But perhaps the most effective road disruptions come from animals that unexpectedly show up outside of their natural habitats. For those drivers in West Bengal, an elephant is no doubt an inconvenience but not a huge surprise. Not so for the Oklahoma couple who were driving home from church and nearly slammed into an escaped circus elephant. "Didn't have time to hit the brakes. The elephant blended in with the road," the driver was quoted as saying. "At the very last second I said 'elephant!' " Fortunately, he had time to swerve, and the couple was unhurt, although their vehicle was punctured by a tusk.

An escaped circus elephant in Zurich, Switzerland, caused a more widespread reaction. The pachyderm spent two hours apparently simply sightseeing, walking calmly around the city and stopping to bathe in a lake, but police closed roads and pedestrians fled.

Smaller exotics can cause just as big a kerfuffle: A wallaby on a motorway in Britain caused a four-hour drama, taking at least nine officers and a police helicopter away from other duties, and causing the speed limit to be lowered to thirty miles per hour.

Similar situations have been caused by llamas in Germany (three police cars with six officers, road closed), circus animals in Dublin (did they head for the Red Cow Roundabout on purpose?), and zebras all over the place: Augsburg, Germany; Atlanta, Georgia; and a suburb of Sacramento, California. A victim of the California incident, whose car was damaged, gives some indication of the determination of these animals on the run:

> I don't know if he went straight into the car or he tried to stop himself and broadsided it. I heard from witnesses he flew up and over my car, got up and kept running.

TRANSPORTATION ALTERNATIVES

Animals interfere with other forms of transportation as well. An elephant in India blocked a train for half an hour and damaged the engine with its trunk before allowing it to continue. Rats have caused dozens of train cancellations and hundreds of delays in England by gnawing on power cables, and to make the rails safe from their rodent dental prowess will cost millions of pounds.

And as if flying weren't unpleasant enough nowadays, animals can make it worse. A stowaway squirrel caused a flight from Tokyo to Dallas to make an unscheduled landing in 2007, and in 2009, a flight in Korea was grounded by the last animal that ought to need to catch a plane: a bird that got into the cabin. Passengers were transferred to another aircraft as the crew tried to catch the lazy freeloading sparrow, resulting in a three-hour delay.

Animals in planes can also cause more than delays. One man in West Virginia flying a private four-seater discovered that he had an unexpected passenger and had to figure out how to make a landing with a four-foot black rat snake wrapped around his arm. And a flight from Houston to Columbus was delayed for an hour when otters got loose from the cargo hold, and weren't just walking around in there:

> A man who had coffee in his suitcase found his bag open and covered in what appeared to be hay.
>
> "Some otters got into them," he said. "They must have smelled the coffee."

FOCUS ON THE BASICS

The roads aren't the only place where animals disrupt our society's infrastructure. Dogs have traditionally interfered with the mail getting through, and cats do their share as well, as we'll see in Chapter 8. But messing with the postman is less effective than it once was, now that so much of our communication is electronic. Unluckily for us, animals have branched out. In 2006, hundreds in Tokyo found themselves without Internet access when crows took a fancy to pecking fiber-optic cable into strips and building nests out of it. Electrical outages have been blamed on cats, squirrels, snakes, and other animals that can climb power lines, including a raccoon described as "very acrobatic and mean-spirited" who caused a short-circuit that knocked out power to a children's hospital and delayed production of the newspaper in Memphis.

The more abstract foundations of our civilization are vulnerable to animals as well: They can even interfere with the democratic process. In India in 2009, rampaging elephants affected voter turnout and overshadowed other political issues in some locales. Said one citizen, "Many villagers did not go out to vote. Those that did had to stand at polling booths fearing that elephants might come by," and another, "This year, our vote will go only to those candidates who will help in getting rid of elephants."

India's animal-political problems have sometimes gone even further than that: In 2007, Delhi's deputy mayor died of head injuries sustained when monkeys attacked him as he was reading the newspaper on his terrace. When he tried to defend himself, he lost his balance and fell.

TREADING ON TRADITION

Animals attempting to undermine our way of life may also target the simple pleasures that we hold most precious.

In Hawaii, young Newell's Shearwater birds inconveniently choose the middle of football season to fly from their nests to the ocean. Every year about thirty of the endangered birds are disoriented by the bright lights of the Kauai stadium and fall to their deaths. The league has been forced to move its high school football games to Saturday afternoons or risk a fine of $30,000 per dead bird.

You might think it's the same game no matter what day it's played, but according to *Hawaii News Now*, this Friday-night tradition is "practically sacred." One parent said, "It's about bringing the community together, it's more than just a football game, it's a way of life." And State Representative Jimmy Tokioka was quoted as saying, "To just go without having football games at night is really going to hurt the social fabric of our community."

Animals may also express criticism of our cultural activities quite directly. In St. Louis, the band Kings of Leon was forced off the stage by pigeons, using the weapon that comes most naturally to them: pigeon poop. Perched in the rafters above the stage, they were apparently most displeased with the bass player, who was hit several times during the first three songs and finally threw in the towel when he was hit in the face. He said, "We had 20 songs on the set list. By the end of the show, I would have been covered from head to toe."

Sometimes, though, it's the little things that really get to you. In a quiet Nottinghamshire village in England, the harassment of a jackdaw has become unbearable, according to one villager:

> He unpegs the washing, turns the pages of the newspaper before you are ready and pick-pockets things from your pocket.

Turns the pages of the newspaper before you are ready. Truly, is nothing sacred?

THREE

Kinky Creatures

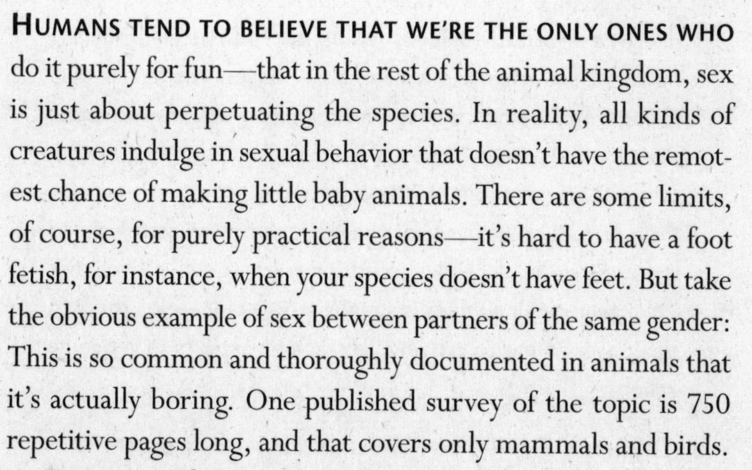

HUMANS TEND TO BELIEVE THAT WE'RE THE ONLY ONES WHO do it purely for fun—that in the rest of the animal kingdom, sex is just about perpetuating the species. In reality, all kinds of creatures indulge in sexual behavior that doesn't have the remotest chance of making little baby animals. There are some limits, of course, for purely practical reasons—it's hard to have a foot fetish, for instance, when your species doesn't have feet. But take the obvious example of sex between partners of the same gender: This is so common and thoroughly documented in animals that it's actually boring. One published survey of the topic is 750 repetitive pages long, and that covers only mammals and birds.

Doing it with someone of the same sex is one way to have fun without the risk of offspring, but as we all know, there's an even simpler way. And animals know it too. After all, who needs a partner when you've got perfectly good flippers of your own?

PROSTITUTION AND PORN

Animals commit other sordid, if less violent, offenses related to sex. Humans may have invented pornography, but animals seem to enjoy it when it's offered. In the lab, monkeys will pay with fruit juice to see pictures of a female monkey's bottom. And in China, keepers have used films of pandas mating to boost the libido of male pandas, who are otherwise notoriously bad at continuing their species; apparently they are particularly aroused by a sexy soundtrack.

Animals came up with prostitution all on their own, though. They'll trade sex for various benefits.

- Fiddler crab females put out for a neighboring male so he'll protect them with his enormous claw.
- Female birds may offer sex to get food, in the form of permission to forage on a male's property. One species of tropical hummingbird vigorously defends its territory of nectar-producing flowers from males and females alike—unless the female performs the courtship ritual and lets him have sex, in which case, she's allowed to feed. This happens outside the breeding season, so again, it's not about producing offspring.
- Male macaques exchange grooming for sex with females, and the price varies depending on the market—and who you are. The lower ranking a male is and the more competition there is, the longer he has to groom a female if he expects to get some action.

If some of those examples seem like less payment than politeness—perhaps it's only civilized to share a nice meal after enjoying each other's company—consider the Adélie penguin. The males may have sex to get their rocks off, but the females have sex to get rocks from them. A female will copulate with a male and then take a stone from his nest as payment, bringing it home to use as material for her own rocky abode.

Reasonable people might disagree about the morality of an honest business of penguin prostitution, but that's not what we have here: They're sneaks and adulterers too. These are married ladies, going off to find single males behind their innocent mates' back. Says the researcher who observed the behavior, "There was no suspicion on the part of the males. Females quite often go off on their own to collect stones, so as far as the males are concerned there is no reason to suspect."

And remember, guys, if her husband can't trust her, she might be lying to you too. Sometimes the female will go through the courtship ritual, then, just when the male thinks he's going to get lucky, she'll grab a stone and run.

Yes, everyone masturbates, and at least animals don't lie about it. It's been documented in many of our fellow primates—at least seventeen species—in the wild and in captivity, and in animals as young as fifteen weeks old. But our close relatives are hardly the only ones: so do kangaroos, walruses, and marine iguanas. And if your parents told you masturbating would make you go blind and it took you years to get rid of the resulting sexual hang-ups, you're not alone. Scientists have shown that punishing stallions for masturbating not only makes them do it more, it interferes with their normal mating behavior. So if you're hoping to breed that next Triple Crown winner, it's best to look the other way.

THE JOY OF FLEX

Aside from doing it with paws or flippers, many animals masturbate using parts of the body—and positions—that you can't.

Did you ever wonder what it would be like to have a tail? Bet you never thought of this use for it, but olive baboons have, as have spider monkeys (their tails are prehensile, which is almost as good as an extra hand). And for male elk, the antlers are erogenous zones—they can rub them against vegetation until they ejaculate.

Some male animals—including squirrel monkeys, marmosets, dwarf cavies, and kangaroos—can reach their own penis with their mouth, with predictable results. And, just to make you even more jealous, fruit bats can copulate and have oral sex at the same time—and scientists say that the simultaneous pleasure helps them do it longer.

Not to be outdone, some female primates can suck their own nipples. Chimps not only get sexual pleasure from this activity—it actually prevents pregnancy by mimicking nursing, which inhibits breeding cycles.

SEX OBJECTS

You might think that with all those body parts and configurations that are unavailable to humans, animals wouldn't bother with sex toys. You'd be wrong.

Our fellow apes commonly use sticks, rocks, and leaves as sexual novelties. And if nothing suitable is available, primates, including orangutans, chimps, and macaques have been seen creating their own. Female orangs, for example, will cut up vines, and males may use a piece of fruit peel or in one imaginative case, a hole poked in a leaf.

Less sophisticated creatures make use of objects for sexual pleasure as well. Both male and female porcupines do it by straddling a stick, and male hummingbirds use small items, like leaves, that they find suspended in spiderwebs. Black-winged stilts, which use objects like bits of driftwood, are particularly enthusiastic—one bird was recorded stimulating himself in bouts of over twenty times in a row twice per minute (a statistic no doubt recorded by a scientist wondering whether his mother was right when she said he should have gone into medicine instead).

ALL TOGETHER NOW

Even when animals have sex in pairs, it's still not always about making babies. Many have sex outside the season when they can breed, and some like it best then—for golden lion tamarins and proboscis monkeys, their peak of sexual activity is actually during pregnancy. And in some animals, like Japanese macaques, the ladies still get plenty of activity after menopause.

Another option that's far too obvious for animals to pass up is oral sex. It's been seen in (of course) primates as well as cheetahs, hedgehogs, and all kinds of fruit bats. Male walruses stimulate each other, as do male manatees.

There is even cross-species oral sex. Caribou and moose often interact, as well as bonobos and red-tail monkeys. There are also reports of grouse and geese and swallows carousing with different species of grouse and geese and swallows, fur seals doing it with sea lions, and gray squirrels getting it on with fox squirrels.

This interspecies sex isn't all heterosexual, either—male chimps will do it with either male or female baboons, male sea otters with both male and female seals, male orangutans with male crab-eating macaques, and male graylag geese with male mute swans.

But none of this should be news, really. Over a hundred years ago, none other than Charles Darwin noted birds mating with birds of other species. He described one case in which a domestic goose had a brood half fathered by her mate and half by a Chinese goose. In another case, a feathered strumpet dumped her mate outright for a bird she preferred:

> Mr. Hewitt states that a wild duck, reared in captivity, "after breeding a couple of seasons with her own mallard, at once shook him off on my placing a male pintail on the water. It was evidently a case of love at first sight, for she swam about the newcomer caressingly, though he appeared evidently alarmed and averse to her overtures of affection. From that hour she forgot her old partner."

It seems that even love can't bridge some cross-species differences, though. Researchers recently observed a seal attempting to mate with a penguin. They considered various possible motives for the seal's behavior, from playfulness to aggression: "At first glimpse, we thought the seal was killing the penguin,"

GETTING IT ON—AND ON AND ON . . .

If animal sex is just for making babies and not for having fun, then why do they do it so darn much? Even when animals are having sex to reproduce, they do it way more than necessary:

- Kestrels and oystercatchers copulate 700 times per clutch of eggs. Tree swallows are more restrained—50 to 70 times is enough for them.
- Female lions mate every fifteen minutes day and night during their four-day estrous period. They may have sex up to 100 times a day during the breeding season, for a total of 1,500 times per litter.
- Female chimps mate an average of 138 times with thirteen different males for each infant they give birth to.

one of them said. But if it was an honest attempt at cross-order romance, he was definitely doing it wrong:

> After 45 minutes the seal gave up, swam into the water and then completely ignored the bird it had just assaulted.

Really, at least send flowers—or some fish—if you're not going to call, you know?

SEX CRIMES OF THE CUTE AND CUDDLY

However nasty the activities we've reviewed so far, perhaps it's none of anyone else's business what goes on between consenting animals. But that kind of nonjudgmental attitude is hard to maintain when there seem to be so many animals that don't get that no means no:

- Up to thirty cute, furry male gray squirrels may chase one female, while she expresses her displeasure by screaming, striking out, and trying to run away when cornered.
- Pretty little hummingbirds engage in high-speed chases where males strike the female in midair and force her down to copulate.
- In some species of geese, half of attempted copulations are not only with someone other than a mate, but are also non-consensual. In snow geese, during some mating seasons each female is subjected to a rape attempt on average every five days. Most—80 percent—of these are directed at females that are not fertilizable, like ones who are incubating eggs.
- Other apparently majestic sea birds are no better. Rape is common in albatrosses: Males will pin a female's neck to the ground or back her into a bush to tangle her up. In one case, a female was observed being raped by four gangs of males in ten minutes, losing an eye and suffering wing injuries.
- Gang rapes among mallard ducks are particularly well documented and quite unambiguous. Commonly, a small flock of males swoops down on a female, displaying none of the normal courtship behaviors. The females resist vigorously, and sometimes they're even drowned. In a particularly charming detail, if she does survive, her own mate typically also rapes her afterward.

One much more primitive animal has developed a particularly despicable strategy that most of us would assume was exclusive to humans: date rape via knockout drugs. The funnel web spider knocks his mate out with a sedative pheromone. After luring her close with an elaborate dance:

> "Suddenly the male will walk forward, and at the same time the female will collapse," says Fred Singer of Radford University in Virginia. Out cold, males take advantage of the senseless female for between several hours and a day. Although she may wake up occasionally, the female is simply re-sedated.

It might be only fair to note that maybe this is the rare case where the male has a legitimate defense, at least if the good of the species is considered:

> The sedative pheromone has probably evolved to defuse females' tendency to eat everything they encounter. Females benefit from this too, says Singer, "if she doesn't [pass] out, she isn't going to mate."

SEX ON THE BEACH

Many of the aquatic mammals that humans find most appealing are in fact the worst sort of sexual predators. Probably the only reason that scientists hadn't seen a seal rape a penguin earlier is that seals are usually much too busy forcing themselves on their own kind: The breeding grounds of various species of seals are patrolled by roaming gangs of males who chase reluctant females around, attacking and raping them if they try to escape.

Other aquatic mammals behave similarly. The offenses of dolphins will be addressed in sordid detail in Chapter 7, but everyone's favorite noble sea mammal, the whale, is no better: Right whales form rape gangs that force a female underwater, after which males take turns mating with her. And even adorable, gentle, vegetarian manatees may form large groups of males and pursue a female for weeks at a time, even when she's not fertilizable. The female and her calf are often injured trying to escape, sometimes with fatal results for the calf.

DON'T KNOW WHEN TO STOP

If you're still hanging on to the belief that the goal of all animal sex is reproduction, consider their most counterproductive strategy of all: Some animals will keep mating with a female till she's dead.

Being aquatic or semiaquatic is clearly a risk factor for this sort of thing. Male frogs commonly pile onto a single female, which may end up drowning her. In some populations of mallards, 10 percent of females die per year due to drowning or other injuries suffered in rapes.

Mating on perfectly dry land is no guarantee of safety, though. Mountain sheep may chase a female for miles, in packs, forcing her to jump to dangerous narrow ledges in an attempt to escape, sometimes with fatal consequences. And even apparently placid domestic sheep reveal these tendencies when deprived of the calming effects of civilization: In one population of feral sheep, males were observed chasing ewes to the point of exhaustion. After they took turns having their way with the ewes for hours, the females ended up battered and too weak to fight off the giant petrels that arrived to finish them off.

It seems pretty obvious that if an animal's goal was to continue the species, this is the worst possible approach. Take the extremely endangered Hawaiian monk seal. The main threat to their survival? Male Hawaiian monk seals, whose sexual attentions often result in females being mobbed to death. Scientists are trying to save them from themselves by rounding them up and giving them drugs to suppress their libido.

It would be nice to think that these cases are all unfortunate accidents, but some male animals really do seem indifferent to whether their mates are alive even at the start of the process: Some will mate with females who are *already* dead. This has been seen in swallows, penguins, and ducks including mallards. In fact, mallards are so undiscriminating, one male was observed mating with a dead male of his own species:

> On 5 June 1995 an adult male mallard (*Anas platyrhynchos*) collided with the glass façade of the Natuurhistorisch Museum Rotterdam and died. Another drake mallard raped the corpse almost continuously for 75 minutes. Then the author disturbed the scene and secured the dead duck. Dissection showed that the rape-victim indeed was of the male sex.

The paper describing this case was awarded an Ig Nobel Prize for biology in 2003, and Dead Duck Day is celebrated on June 5 every year at the Natural History Museum of Rotterdam to honor the memory of the victim of this first scientifically documented case of homosexual necrophilia in the mallard duck.

KEEP YOUR PAWS TO YOURSELVES!

This would be none of our business if animals kept their perversion to themselves. But they often make their sex lives our problem, as anyone knows who's ever tried to explain to a child at the zoo just what that groaning pair of tortoises is up to.

And it's not just that we have to watch them doing it to each other. Remember those cross-species sexual attractions? They can be directed at humans as well. Apes are frequent offenders, and they can have particular tastes—one chimp at a zoo in Florida not only preferred blonds but had a fetish for shoulders, strutting to impress lady visitors in tank tops and cajoling the female keepers to show him a little skin.

It's not just those big-brained creatures, either. One kangaroo stalked women of the perhaps significantly named town of Honeymoon Ridge in Australia. A victim who'd been walking along a bike trail thought nothing of it at first when she saw a big kangaroo behind her:

> "I turned around and saw this big kangaroo behind me, so I hastened my steps," she said. "It seemed a bit odd, but I continued walking and didn't think much about it."
>
> "Then on the return walk he was there waiting for me," she said. "With his male pride on full alert, he started circling

> me. There was no doubt about what he wanted, the randy old thing."

Fortunately, the animal retreated when other walkers appeared on the trail, but he was not too intimidated to later approach a woman in a crowd of spectators at a race. She was oblivious to his intentions, but they were quite clear to eyewitnesses: "Yeah, apparently he was quite aroused," said the object of his attention. "I'm actually glad I didn't notice."

And even if humans aren't the target of a creature's warped desires, an innocent bystander can be held hostage to them. In Witt, Illinois, a woman was trapped in her house for several hours when a gander tried to mate with her lawn ornament, a concrete goose. Every time Joanne Martin tried to open her front door, the lust-crazed bird attacked her. Eventually six men tried to assist, without success.

> "Brent Bourke beat the heck out of him with a stick," says Joanne, "and ran him out on Highway 16. . . . He was going pretty good, but then the goose turned around and came back at him. That was the funniest part."

Bourke was chased into the house, as was another man who tried the same method.

The goose was even undeterred from its mating frenzy when firecrackers were thrown at it. Finally, in a concerted effort, a couple of the men held the goose at bay long enough for the others to grab the statue and close it in a shed. With the object of his passion now inaccessible, the frustrated suitor eventually wandered off. The story was over . . . or was it? A reporter visiting a few weeks later to photograph the concrete goose made an ominous observation:

> As I pulled away, the goose safely back in the shed, I noticed it wasn't Joanne's only lawn ornament. She has two concrete deer.

FOUR

Animal Family Values

Whether you blame nature or nurture or a little of both, bad behavior has to come from somewhere. As earlier chapters have shown, animals are often criminals and perverts, so it should come as no surprise that most of them have dysfunctional families as well.

Much is made of the maternal instinct, but good parenting hardly seems to come naturally in nature; in fact, some animals don't bother to parent at all. Frogs, for example, deposit thousands of eggs, figure that at least one will survive on its own, and get on with their lives. Many egg-laying animals do the same, even some birds: The scrub fowl gets out of staying home to keep the kids warm by burying its eggs in a pile of rotting compost.

And sometimes the abandoned babies are the lucky ones. It's better than the alternatives: The same devoted mother hamster that licks her pups clean and suckles and protects them will often chow down on a couple of the tender morsels as well.

There was a time when scientists thought that only humans killed members of their own species. Our noble fellow creatures killed only for food, and they certainly didn't murder their own babies. Those were the good old days. However, in the 1970s, researchers finally admitted that in many species, infanticide—and sometime cannibalism—are, as one put it, "everyday occurrences" in the animal world.

Some offspring are killed by neglect. Lionesses may abandon litters when food is scarce. Flamingo colonies will skip breeding for three or four years at a time, and when they do decide to make the effort, up to half of the pairs may change their minds and abandon their eggs. Oystercatchers often starve their chicks out of what is apparently sheer laziness. Those with nests farther inland, instead of right next door to their feeding areas, can find plenty of food, but apparently they just can't be bothered to transport it to their chicks, even though researchers have found that they spend 37 percent of their time doing nothing.

In many cases, however, the perpetrators of infanticide are a bit more active. Kangaroos may dump a joey when chased by a predator. In Australian sea lions, the most frequent cause of death to babies while on land is attacks by adult female Australian sea lions; other species of seals and sea lions are no better. And being eaten by adults, even by one's own parents, causes up to a quarter of chick mortality in herring gulls—in one study, three hundred such deaths were counted in one colony.

SOCIAL CLIMBING

With many social animals, the mother's status in the community is more than a matter of popularity—it's actually a matter of life and death. Dominant female chimps will kill babies born to other females, and in those charming, cooperative extended families of meerkats, pregnant females will often kill other females' litters. And in both cases, the babies aren't just killed—they're eaten, too.

Not all the blame falls on the ladies by a long shot, though. In many species, when a new male moves in, the first thing he does is kill the existing babies. When a cute little house sparrow who has lost her mate gets a new husband, he pecks her young to death to make room for his own. And female lions have an apparently heartwarming method of raising their cubs in communal nurseries; turns out, this isn't because lions have socialist utopian childrearing ideals, but because lionesses that raise their cubs together can better defend them from bloodthirsty newcomers looking for mates.

Of course, the females' reactions to these murderous males doesn't exactly discourage them. What do the bereaved mothers do when males kill their offspring? They often have sex with them, and not unwillingly: It even happens in the very rare species (like langur monkeys) in which rape is unknown.

COMING BY IT HONESTLY

Baby animals take after their parents from the very start: they kill their fellow babies too. Some don't even wait to be born: Pronghorn antelopes may grow a kind of spike in the womb that kills other embryos, and sand tiger shark embryos actually attack and eat one other. Hyena cubs fight with their siblings from the moment of birth, often killing them. Piglets do the same, inspiring the authors of one scientific paper to write the abstract in poetry:

A piglet's most precious possession
Is the teat that he fattens his flesh on.
He fights for his teat with tenacity
Against any sibling's audacity.
The piglet, to arm for this mission,
Is born with a warlike dentition
Of eight tiny tusks, sharp as sabres,
Which help in impressing the neighbors.

The champion sibling killers, though, are birds. In many species, statistically the biggest risk to a youngster's survival isn't predation, starvation, or disease—it's a brother or sister:

- Blue-footed booby chicks will push a sib outside the white circle of guano around the nest that the parents defend as their territory. The adults don't allow anything to enter the ring—even their own displaced offspring—so the chick can't get back into the nest and dies.
- Kittiwakes will push a sibling out of their precarious cliffside home, and even if it doesn't instantly plunge to its death, once it's out of the nest, the parents ignore its cries for help.

This behavior is not a rarity: in fact, some species practice what scientists call "obligate siblicide." This dry technical term means that the parents usually lay two eggs, but the resulting offspring make sure that only one of them survives.

This sort of excessive sibling rivalry might be excused when there isn't enough food to go around, but in many cases, that's not the problem. In one case, a black eagle chick was reported to have ignored several delicious hyrax carcasses to keep beating up on its sibling, which it pecked more than 1,569 times in 187 minutes of fighting.

And, like the parents that ignore the cries of chicks that have been sneakily pushed out of the nest, the adults are equally unhelpful in the face of this violence. As one researcher reports, they seem to think they have better things to do:

> During sib-fighting of great egrets in Texas, the attending parent typically preens or loafs, frequently not even watching as the victim is pummeled.

MOMMIE AND DADDIE DEAREST

If infants manage not to get eaten, starved to death, or murdered by their siblings, that doesn't mean they're in the clear. There's still the risk of child abuse by parents as well as by other adults who don't seem to believe that it takes a village to raise a baby animal:

- Ever been driven crazy by a crying child in a public place? Rhesus monkeys feel the same way, and they're not very nice about it. Mothers rush to give in to a temper tantrum when other monkeys are around, because otherwise the onlookers will kick, bite, and shove the mom and her annoying tot. The reason is familiar, according to a researcher: "They do it basically because they are annoyed by the sound."

- Ring-billed gulls incestuously mount their own chicks—and it's mostly females who do it, some of which are habitual offenders.
- Herring gulls mount their chicks too—the ones that manage to be born in spite of the fact that they also eat their own eggs. And if youngsters try to leave home to escape abuse, they don't find help elsewhere: The youngsters are pecked, picked up and shaken and thrown, and killed and occasionally cannibalized by other adults.
- In some species of monkeys, a certain percentage of mothers are serial child abusers. Moms in one colony of pigtail macaques were observed taking normal care of the kids as well as physically abusing them—dragging them by the tail or legs, pressing them against the ground, chewing on their extremities, and compulsively grooming around the eyes, sometimes causing blindness. And on an uncannily familiar note, researchers have discovered that it runs in families: If you're a macaque raised by an abusive mother you're more likely to be an abuser yourself.

Animal parents play favorites, too, and sometimes for the lowest of reasons: Earwigs take better care of offspring that are already well fed and vigorous and neglect the hungry ones that really need their care. And coots give more food to the chicks that have the most bizarrely colorful plumage.

SCREWING AROUND

Perhaps it's only to be expected that animals are bad parents—it's what happens when you bring children into a relationship that's already a mess. Even putting aside incest and other warped lifestyle choices (swans, for example, are happy to marry their parents or siblings), our fellow creatures can't seem to manage simple monogamy.

We used to believe that many animals formed faithful, till-death-do-us-part couples. In reality, they're so good at sneaking around that they even fooled the experts. For instance, birds have traditionally been held up as paragons of monogamy, so it was a shock to scientists when a project to control blackbird populations failed. The paired-up guys were all vasectomized and released—but somehow, all their mates had offspring anyway. Now that DNA analysis is possible, what's clinically called "extra-pair paternity" has been proven in birds all over the world, in every family. In the most promiscuous bird found so far, the salt marsh sparrow, 95 percent of females mate with more than one male. Researchers in Connecticut found that nearly every clutch of eggs had more than one father—an average of 2.5 daddies per nest—and in a third of the nests, every chick had a different father. Even a scientist hardened to the truth about bird relationships was taken aback:

> "We were not surprised to find some level of promiscuity," he said. "But we were quite stunned at just how extreme the rate was."

MISSING THE POINT

When humans take an interest in animal family life, they tend to get the issues exactly backward. For example, staff at a garden in Somerset, England, must constantly reassure visitors who worry about a male swan living as a bachelor. "The public do get upset when they see the swan by himself because the general impression tends to be of swans being part of a loyal couple, devoted to one another," says the head of the gardens.

However, the reason the swan is alone is that he can't be trusted around the ladies—in fact, he's possibly a serial wife murderer. He's suspected of killing his first mate, who was found dead in the water while incubating eggs. The second apparently fled with the children, and the third "didn't want to spend any time with him and later died of 'depression,'" according to news reports.

In another demonstration of our misconceptions about bird families, many animal fans have been charmed by the devoted relationship between the adults portrayed in the picture book *And Tango Makes Three*, which depicts a gay penguin couple raising an adopted chick. However, the story bears little resemblance to reality. Divorce rates in various species of penguins are close to 50 percent, and promiscuity is rampant. In one species, one third to one half of heterosexual activity is adulterous, and nearly half of the gay sex involves male mates getting it on the side.

Penguin parenting is also a less rosy story in real life. Scientists have shown that penguins won't work harder to pick up the slack for a handicapped spouse: When one Adélie penguin of a pair was fitted with a tracking device that slowed down its swimming, the other parent didn't bring back more food for the chicks to compensate.

And gay penguin parents are no exception to these general trends. In fact, another long-term same-sex penguin couple in San Francisco who'd also raised an adopted chick broke up when one of them left his partner—and for a female penguin, no less.

FUZZY INFIDELITY

The behavior of birds may be appalling, but don't be too quick to feel morally superior just because you're a mammal. Unlike birds, who at least maintain a facade in front of the children, less than 3 percent of mammal species even bother to pretend to be monogamous.

And that exceptional 3 percent? Turns out they're the ones who are exceptionally good at hiding their affairs from the prying eyes of researchers. Prairie voles are supposedly part of that mammal minority that spend their whole lives with a single partner, and scientists have flocked to study this phenomenon of furry fidelity. However, it turns out that these models of monogamy routinely cheat. The females may share a home with only one male, but they'll share a bed with others: A DNA study showed that almost a quarter of litters had a father other than the live-in mate. The hardened author of the study, apparently no longer capable of being shocked by the sex lives of animals, said:

> You can pair with a partner for life and still have sex with others—and that is what prairie voles do. There is a lesson there for humans.

This disappointing news about rodent morality may help solve an abiding mystery of animal family life, though. Some species live in groups where young adults, rather than leaving home to start their own families, help raise the offspring of the dominant pair. Evolutionary theorists have twisted themselves into knots trying to explain how such a social system can persist when the assistants don't get to pass on the genes for their accommodating behavior. But part of the answer may be simple: The "helpful" male, at least, may not be so selfless after all. Chances are that he's having sex with the group's mom too, and some of the kids he's helping raise are actually his own.

FAMILIAR DRAMAS

Many family dramas are repetitions of sad old stories—and sound oddly familiar. In recent years in some parts of Africa, large numbers of rhinos were being sexually assaulted and murdered. The mysterious deaths were eventually traced to young male elephants, who were also killing each other at unprecedented rates: In one park, up to 90 percent of male elephant deaths were due to other elephants.

What's to blame for this heightened aggression? It turned out that these violent youth gangs were made up of the sons of single teen mothers. Because of human culling and relocation of herds, inexperienced young females were raising babies without the support of grandmothers and other female relatives, and their sons were growing up without adult male role models. The solution, it turned out, was support for creating traditional families: In one park, when rangers moved in older bull elephants to join the group, the young studs were brought under control.

FIVE

Party Animals

SUBSTANCE ABUSE HAS TO BE THE ONE KIND OF BAD BEHAVIOR that's unique to humans, right? Doing drugs, getting drunk—surely that's not possible without the technology that gives us the ability to drive to the liquor store and pay for a six-pack, or at the very least, without opposable thumbs for opening bottles?

Sadly, this is completely untrue. As fermentation is a natural process, alcohol does occur in nature, and it's hardly just the highly evolved that have a taste for the stuff. Animals including cows, goats, pigs, and monkeys are happy to chow down on fermented fruit. In yet another example of how wrong our romantic notions of nature are, one of the best documented animal drunkards is one of the few insects that humans have a positive feeling about: the honeybee.

BOOZING BEES

The drinking problems of bees have landed them in the news for as long as there's been news, as seen in this 1898 *New York Times* report about bees hanging out around sugarcane factories:

> At first the bees carry on their labors diligently. Then, little by little, they learn that juices from the sugar cane contain alcohol. . . .
>
> Forsaking even the semblance of work, the bees imbibe the intoxicating fluid, and thenceforth the social and mental decline is marked. The sad fact is that the bees get drunk. They fly about in a dazed and listless condition, ambitionless so far as honey making is concerned. Once they have drunk from the fountain of Bacchus, they are moral and physical degenerates.

Some scientists take advantage of this taste for booze in the lab, using bees in alcoholism research. They're the perfect subject, says one researcher:

> Most animals have to be tricked into drinking alcohol, says Charles Abramson of Ohio State University. But a honeybee will happily drink the equivalent of a human downing 10 litres of wine at one sitting.
>
> "We can get them to drink pure ethanol, and I know of no organism that drinks pure ethanol—not even a college student," he says.

CAN AN APE GET SOME SERVICE HERE?

Lacking either grant-funded libations or the chance to snatch insufficiently supervised beverages, some animals will demand to be served by humans in any way they can. One chimp in Russia named Zhora was actually removed from his zoo and sent to rehab to treat his smoking and drinking habit. Some reports implied that this tragedy was all the fault of zoo visitors who, despite the pleas of management and a barrier of three fences, managed to supply the chimp with booze and cigarettes. But they were only giving Zhora what he asked for—as Reuters quotes the zoo director: "He would pester passers-by for booze."

The results of their enthusiastic participation in this research confirm the observations of the nineteenth-century *Times* reporter. Bees share with us not just the taste for liquor but its effects as well, including memory loss, impaired motor functioning, and poor personal hygiene:

> Researchers gave honey bees various levels of ethanol, the intoxicating agent in liquor, and monitored the ensuing behavioral effects of the drink—specifically how much time the bees spent flying, walking, standing still, grooming and flat on their backs, so drunk they couldn't stand up. . . .
>
> Not surprisingly, increasing ethanol consumption meant bees spent less time flying, walking and grooming, and more time upside down.

GIN AND TONIC, PLEASE

Even when given alternatives, many animals choose alcohol. Scientists devised a fruit fly–size drink dispenser (no word on whether they also invented tiny fruit fly–size paper umbrellas) and found that given the choice between plain and alcoholic beverages, fruit flies preferred booze. They even developed a tolerance, gradually coming to prefer stronger drink. The researchers also observed drunken behavior, although in fruit flies, this was pretty much confined to "hyperactivity and loss of coordination," since the flies were not given access to lamp-shades to put on their heads or cars to drive into stationary objects.

A side result of this research was also intriguing, if you've ever wondered how certain alcoholic beverages become popular and traditional despite how nasty they taste. When quinine was added to their drinks, the fruit flies, which usually avoid the toxic, foul-tasting stuff, thought it was just fine as a mixer. "I was actually pretty surprised when they continued to drink it," one researcher said.

CHOOSE YOUR POISON

You may be thinking to yourself, *These are probably just hungry animals who have stumbled across some rotten fruit. An animal's got to eat, right?* Well, scientists are way ahead of you. A study by psychopharmacologist Ronald Siegel (who we'll be hearing a lot more from in this chapter) found that hunger had nothing to do with it. In a barn with free access to food, water, and

REINDEER FLYING HIGH

Animals also don't miss out on the magic of mushrooms. Reindeer in Siberia not only eat the fly agaric mushroom and then behave like they're stoned but will even consume the urine of human mushroom eaters to get the same effect. Herders take advantage of this by using urine to assist in reindeer roundup, but the animals can be so aggressive about getting their fix that travelers are warned against peeing within sight of one.

Dogs go for psychedelic 'shrooms, too, and researcher Ronald Siegel, who's made a career of testing and observing animals on drugs, reports:

> *On ranches in Hawaii and Mexico, I saw dogs deliberately nipping the caps off psilocybin mushrooms and swallowing them. A few minutes later the dogs were running around in circles, head-twitching, yelping and refusing to respond to human commands.*

alcohol, elephants chose to drink the equivalent of about thirty-five cans of beer a day. The effects were noticeable:

> They started growling—a vocalization pattern associated with arousal—and flapping their ears more than usual. . . . They began swaying rapidly for an hour or two, then slowed down and leaned against their chains, which prevented them from falling over.

And just like us, drink kept these elephants from doing their jobs:

> Flapping their ears like a frustrated Dumbo, they had difficulty responding to commands from their handler. They kept dropping their trunkhold on each other's tails—the trained elephant's version of walking a straight line.

Siegel also tried offering alcohol to a free-ranging herd in a game park. He found that being bartender to elephants is not without its risks: They not only fought among themselves for access to the vat of alcohol, but when he tried to cut one off, it pursued his jeep and attacked him.

In fact, when not being plied with free drinks by generous researchers, animals will make special efforts to get their beverage of choice. Vervet monkeys on the Caribbean island of St. Kitts developed a taste for liquor by eating fermented sugarcane in cane fields, but they've long since found a better source and have become famous for stealing vacationers' unattended drinks. In a Washington State campground, a bear was found sleeping off thirty-six cans of beer, but he clearly had a favorite brand. Evidently, the bear had tried one can of Busch and rejected it for the local favorite Rainier. He got his comeuppance, though, when wildlife officials took the cue:

> The agents decided to trap the bear with doughnuts, honey and, of course, two cans of Rainier beer. It did the trick and he was captured.

In India, elephants commonly break into fermenting vats of rice beer, and it's not because they stumble on it accidentally. "It has been noticed that elephants have developed a taste for rice beer and local liquor and they always look for it when they invade villages," an elephant expert in Guwahati told reporters. In southeast Asia, pigs and chickens are reported to share this taste as well.

THE MORNING AFTER

It's clear, then, that animal drunkenness is not something that only happens when wicked scientists and zoo visitors push strong drink on innocent creatures or when nature or inattentive vacationers leave it where curious critters can stumble across it. Boozing beasts may have their reasons. It's often been suggested that elephants, weary of their renowned perfect memories, drink to forget. But before you feel sorry for our fellow imbibers, remember that animal drunkenness isn't always a victimless crime.

Sometimes the offenses are fairly minor, like the badger, dead drunk on fermented cherries, who was blocking traffic in the middle of a road in Germany. Others are more serious. Indian elephants drunk on local rice beer have brawls that can be fatal both to humans and to the pachyderms themselves, like the ones that electrocuted themselves on power lines.

In one strange case, a man in Sweden was arrested and jailed for ten days after finding the body of his wife, who had not returned from walking her dog. He was cleared only after technicians found physical evidence on her clothing pointing to a different perpetrator: hair and saliva of an elk—a normally reclusive animal that can become aggressive after eating fermented fallen apples in gardens.

And like humans, animal drinkers persist despite suffering unpleasant aftereffects themselves. Darwin wrote about an ancient method of trapping monkeys that involved leaving out open jars of wine, which the monkeys would drink till they passed out. He observed that they then woke up with hangovers:

> On the following morning they were very cross and dismal; they held their aching heads with both hands, and wore a most pitiable expression: when beer or wine was offered them, they turned away with disgust, but relished the juice of lemons.

Modern scientists studying the monkeys of St. Kitts observe parallels among partying primates:

> The parallels between the vervets' behaviour and human behaviour are striking. A cageful of drunken monkeys is like a cocktail party. You have one who gets aggressive, one who gets sexy, one who thinks everything's funny and one who gets really grumpy.

In fact, researchers have found strangely familiar drinking patterns in general. The monkeys can be divided into the same categories as humans—there are binge drinkers, steady drinkers, social drinkers, and teetotalers. The majority drink in moderation and prefer their alcohol mixed with something sweet, but there are small percentages who never drink or who go to the other extreme:

> The binge drinkers gulp down the alcohol at a very fast rate and pass out on the floor. The next day they do it all over again.

But while most animals pay the price of overindulgence, some get off completely scot-free. A certain tree shrew in Malaysia lives on a diet of fermented nectar, and scientists found that the shrews showed no signs of intoxication despite blood alcohol levels that would be several times the legal limit in most places.

And next time you're planning your Saturday night transportation, take a moment to be envious of fruit bats. They don't just have better sex than you, as we saw in Chapter 3, but they don't need a designated driver, either. Researchers who fed alcohol to fruit bats in Belize and then had them fly through an obstacle course predicted that they would basically stumble through it, colliding with stationary objects. Contrary to expectations, the bats' flight and navigation ability was completely unimpaired.

BEASTS BAKED, BLASTED, AND BUZZED

Strong drink isn't the only mind-altering substance that our fellow creatures indulge in. Many of us who live with cats frequently supply them with their drug of choice, either straight or in cute little toys. Catnip is no use for other species, but like alcohol, other drugs can affect our fellow creatures in familiar ways. Here's the owner of a dog that ate a stash of pot found in a park in Seattle:

> "His eyes were kind of glossed over, very out of touch, I mean, he didn't seem to recognize me at first," Nestor Waddell said. "When he was trying to walk, he was looking at his

paw, and then looking at the ground and then trying to get his paw to reach the ground, but was unsuccessful."

Just like with booze, recognizable behavior when high on drugs isn't confined to creatures that are our close relatives. Our friends the bees, when given cocaine, dance more—and remember, in their world dancing isn't just about partying, it's about communication. The wiggle dance of the honeybee provides hive mates with directions to a food source, and it's normally calibrated to the quality of the food and how badly the hive needs it. But when the bees are coked-up, they tend to exaggerate: They're more likely to dance, regardless of those factors.

YOU CAN LEAD A HORSE TO LOCOWEED . . .

While at first glance it may seem a bit unfair to blame animals for what happens in a lab, scientists know better. Just like in the alcohol studies, they've gone out of their way to design experiments that give animals a choice. These days, study animals can even self-administer a dose by pushing a button or lever attached to an injection pump.

But the first experimenters back in the nineteenth century didn't need fancy medical technology to observe the same effect. Siegel describes what happened in the earliest studies on morphine and opium: Pigeons "came to the front of their cages, wings flapping with excitement, and stood without protest for their next injection," and cats, after initially trying to bite and scratch, "would run to the experimenter, jump on his lap, and even lick his hand while waiting for the morphine."

And it's not just scientists who help animals develop a habit. In some parts of the world, opium is still used as a training reward for elephants. This may seem fiendishly clever—who's more motivated than an addict looking for a fix? However, they can apparently become rather insistent on the matter. Imagine a jittery, drug-deprived, four-ton beast searching your pockets with its trunk, and you might decide it's safer to stick to peanuts.

In any case, animals don't need human assistance to develop a craving for mind-altering substances. Locoweed, which grows in the American West, was first described by an early settler who observed his horses having hallucinations and fits. Far from learning to avoid the plant, livestock will return to it insistently, and when farmers work to clear it from the fields, the animals will try to steal it back.

In addition to making use of drugs they find in the wild, animals will steal our own hard-gotten harvests whenever they can. Siegel reports that dogs and goats will swipe any peyote left within reach by their Indian owners, and describes a trio of goats that ate a large stash and spent the day running amok, charging and butting each other and humans as well, occasionally pausing to stare into space and twitch their heads. (Admittedly, goats probably wouldn't have spent that day doing anything much more productive anyway.)

POPPIES . . . POPPIES . . .

In the nineteenth century, animals enthusiastically participated in the fashion for opium, like the two blackbirds in one opium den who would perch near a human user and share the smoke. Writer Jean Cocteau observed:

> All animals are charmed by opium. Addicts in the colonies know the danger of this bait for wild beasts and reptiles. Flies gather round the tray and dream, the lizards with their little mittens swoon on the ceiling above the lamp and wait for the night, mice come close and nibble the dross. . . . The cockroaches and the spiders form a circle in ecstasy.

While most humans shun opium in favor of more modern mind alterants, some animals still find a way to get a fix. In Australia, wallabies have been caught eating opium poppies and getting, in the words of a government official, "as high as a kite." The effects of the opium inspire them to hop round in circles in the fields, creating crop circles (and frightening human bystanders who fear an alien invasion). Sheep are also reportedly developing a taste for the poppies.

However, there's still at least one place where animals can indulge in human-processed opium, and on a grander scale. An opium factory in the northern Indian state of Uttar Pradesh has been producing the drug from poppies since 1820, and one thing that hasn't changed is that humans aren't the only ones who take advantage of it, as the BBC reported:

> Monkeys still have the run of the factory, eating opium waste and dozing all day.
>
> "They have become addicted to opium. Most of the time we have to drag dozing monkeys away from this place," a worker says.

ANIMAL PUSHERS

In fact, far from humans pushing drugs on animals, it's sometimes the other way around. Animals are to blame for introducing us to many intoxicants. Khat (also spelled *qat*), which is chewed as a stimulant in East Africa and the Middle East, is said to have been discovered when a herder noticed that his goats were wired after eating the leaves and decided to follow their example. (One source describes the effect of qat as producing "feelings of euphoria and alertness that can verge on mania and hyperactivity," which is about the last thing you want in a goat.) A researcher who studied the use of various drugs by African apes notes that they introduced humans to a hallucinogenic root used in Gabon:

> Most intriguing, said Professor Huffman, is how local people claim to have discovered the intoxicating effects of the plant by watching animals, including gorillas, go into a frenzy of fear, as if being chased by invisible objects, after eating the roots.

Why this observation didn't serve to warn people away rather than attract them is a subject for another book entirely.

We don't even need to go so far afield to find a case in which animals are to blame for our own addictions. Legend has it that we were introduced to coffee by goats. The story is similar to that of khat: A herder observed that his (no doubt already obnoxious) goats were even more energetic than normal, and realized that this was the result when they ate the berries of a particular shrub.

The most charming renditions of the tale describe the goats "dancing" on their hind legs and the young goatherd's concern that he would get in trouble if his herd became ill from eating the strange fruit. However, there's also a hint of something darker: One character who's particularly intrigued by the goats' behavior is the head of a monastery, whose monks "had considerable difficulty keeping awake during their nocturnal devotions."

Have you ever suspected that coffee was deliberately invented to keep you working for hours on end when a reasonable person would take a siesta, demand an assistant, or quit their job? That it is, in fact, a tool of The Man? Well, now you know who to blame for providing it to The Man: It's The Goat.

HYPNO-TOAD AND COMPANY

All the animals we've seen so far have confined their abuse to, well, *normal* drugs. However animals don't restrict themselves to just plants and fungi to get a high—they'll even use other animals. Certain primates, like lemurs, will catch a millipede, give it a little bite to stimulate its defensive toxins, and rub it all over their bodies. The toxins do seem to repel insects and parasites, but the lemurs' half-closed eyes, drooling, and dazed dozing-off clearly shows that they're getting another benefit as well.

Toad-licking to get high may be an urban legend as far as humans go—unless you count people who heard the tale and were stupid enough to give it a try—but apparently it's such a

THIS IS YOUR CAT ON CATNIP

You don't need to apply for research grants or go on an expedition to an exotic wild locale to observe animal drug use in a natural setting—all that's necessary is to plant some catnip in your own yard. Owners describe drooling, dilated pupils, and aggressive defense of the stash as some of the reactions of their cats to catnip. And much of the typical behavior suggests that cats are actually hallucinating: gazing into space at things that aren't there and batting at birds and pouncing on mice nobody else can see.

That's all bad enough, of course. But here's an odd fact: Catnip does not work on kittens. It doesn't affect cats till they get older. Want to guess why?

Cats react to catnip only once they have reached sexual maturity because its active ingredient is basically an artificial feline sex pheromone. The rubbing, vocalizing, and particularly the rolling-around are the same behaviors we see in female cats in heat.

big problem for canines that in Arizona, where the native toads have a lethal toxin, you can sign your dog up for toad-avoidance training classes.

In Australia, some dogs seem to have learned how to get the benefits despite the danger:

> Megan Pickering, a vet in Katherine, said she had treated a number of dogs affected by the deadly toad poison.

"We have had quite a number of cases of dogs that are getting addicted to the toxin," Ms Pickering told the *Northern Territory News* newspaper.

"There seems to be dogs that are licking the toxin to get high. They lick the toads and only take in a small amount of the poison—they get a smile on their face and look like they are going to wander off into the sunset."

JUNK FOOD JUNKIES

Maybe the biggest surprise in the annals of animal addiction, though, comes from the scientists who proved that they can be junk food junkies. Anyone who's seen rats foraging in Dumpsters would probably suspect that they are attracted to the worst products of humankind's fake-food factories, but the effects are way more profound than you might expect. When two neuroscientists fed rats a diet of junk food including Ho Hos, sausage, pound cake, bacon, and cheesecake, it was no surprise that the rodents became obese. But they also became compulsive eaters, and the researchers found that their brains showed the same changes as those of addicts. The rats continued to eat junk food even if they learned that an electric shock would follow it. And they find it just as hard to break the junk food habit as you do:

> When the junk food was taken away and the rats had access only to nutritious chow (what Kenny calls the "salad option"), the obese rats refused to eat. "They starve themselves for two weeks afterward," Kenny says. "Their dietary preferences are dramatically shifted."

Maybe this should be the basis for the next big government nutrition education campaign. The next time you reach for that bag of chips, think about it—Do you really want to lower yourself to the level of a rat?

SIX

Beastly Devices and Deceits

IT'S COMFORTING TO ASSUME THAT ANIMALS ARE ESSENTIALLY honest. After all, when your dog comes running to greet you at the end of a long day, no one wants to have to wonder if he's faking it. And if there's a moment of doubt, we can remind ourselves that it takes a quick wit to come up with a good cover story. Surely they're just not smart enough to tell anything besides the truth.

Sadly, this reassuring rationalization doesn't stand up to the facts. Lying goes back way in evolutionary history, maybe because it's just too useful for nature to miss out on. For many animals, in fact, deception is part of their fundamental biology. There are insects that look like another insect with a nasty flavor, relieving them of the burden of actually producing their own bad-tasting bodily fluids. Some harmless snakes imitate the appearance of venomous species, and don't even have to do that good a job of it, since most animals (including us) can't remember whether it's red next to black or red next to yellow that's the poisonous one.

There are more complicated versions of this kind of physical deception as well. In some species of fish there are what even scientists call *sneaker males*—by staying small and looking like females, they can sneak into a nest and fertilize eggs right under the nose of the dominant male. In the flat lizard, "she-males" keep their juvenile coloration for the same reason—the dominant guy won't chase them off, so they get a chance to mate with his girlfriends. (They still smell like males, though, so they need to be prepared to get a move on if he gets too close.)

If you can't take those cold-blooded creatures seriously, wait: This kind of thing also happens in a much more advanced species—a close relative, even. In orangutans, the dominant males are big bulky brutes, hairier and with wider faces than youngsters. But there are also adult males—the "Peter Pan" type—that keep looking like adolescents, sometimes for as long as twenty years. This lets them sneak around and mate with females without being driven away by the head honcho—and if the big guy meets with an unfortunate accident and Peter Pan gets to take over, soon he will be pumped up and bearded as well.

THE LYING GAME

It might not seem fair to accuse those animals of lying. No one gets to choose the color of their complexion, right?

Ah, but some animals do. And guess what they do with it?

Cuttlefish can control the color and pattern of their skin, changing it with amazing speed and great variety. And in the giant Australian cuttlefish, sneaker males use this ability to cross-dress. They will hide their extra male arms and hold the others up the way an egg-laying female does, but that's not enough to fool the sharp vision of another cuttlefish. So they also change their skin to the typical mottled female coloration. "They are actually disguising themselves to get past the males they couldn't beat in a fight," says one researcher. Their tranny disguise can be almost too successful, causing all kinds of cephalopod sexual confusion: Sometimes other males will hit on the fakers, since even some sneaker males fail to recognize their fellow sneaks.

PLAYING POSSUM

Few animals have the quick-change talents of the cuttlefish, but less fancy changes to your appearance can work just as well, like when a bird pretends to have a broken wing to distract predators from her nest. And while that behavior may be a simple instinct, there's also the chimp who was injured in a fight with a rival and limped for a week afterward—but only when his competitor was watching.

Animals also take it further and look not just injured, but dead. And it turns out that playing possum to fool a predator is not as innocent a strategy as it might seem. Scientists have studied beetles that are preyed on by spiders to see exactly how this ruse works: Is it just that the predator loses interest? They found that in fact, playing dead works okay when you're alone, but it's most effective when the other creatures around you react by fleeing. The spiders go after them instead, so the beetle who fakes it survives by sacrificing its neighbors.

HEY, LOOK OVER THERE!

Appearances may speak volumes, but actual communication systems can be pretty handy too. And it seems like as soon as a species can communicate with sound, lying is quick to follow.

Many animals use alarm calls to alert their fellows that a predator has been spotted. Some even have different calls for different kinds of threats, so you know whether you're fleeing an eagle or a leopard and can use the appropriate strategy to escape. Helpful, cooperative behavior, right?

Of course, the whole point of this sort of thing is to react immediately, without wasting precious moments second-guessing—which makes it easy to use for other purposes. When you and another bird see a tasty insect at the same time, and the other guy says "Whoa, an eagle!" you're going to look up automatically—while your helpful friend, knowing there's no danger, snatches the treat from under your beak.

If there's too much crying wolf, though, everyone's at risk of ignoring a real danger. So animals don't waste this clever strategy if mere bullying will suffice. For example, great tits use alarm calls to chase a dominant bird away from food, but if it's a subordinate, they threaten to beat it up instead.

Some animals have even figured out how to use this ruse on different species: A South African bird called the drongo imitates the alarm calls of other birds. Their fake calls chase those birds away from food, and also fool meerkats, who've learned to rely on the birdcalls for their own safety. (And although it hasn't yet been proven to the picky satisfaction of science, the drongo has been observed imitating the meerkats' own predator alarms as well.)

BORN TO BE FOOLED

Apes may be able to manipulate humans with tools, enticing us with a piece of straw and circumventing our security measures with a little bit of wire, but our own pets don't need any additional technology. They just have to use our own biology against us:

- Those puppy-dog eyes that make you forgive your dog for bad behavior trigger the same neurotransmitter involved in pair-bonding with your mate and maternal behavior toward your offspring.
- Cats are much better at ordering us around than vice-versa: There's a special combination meow–purr some cats use to demand service, and it's acoustically similar to the cry of a human infant. We perceive it as less annoying than a meow—so the cat doesn't get kicked out of the bedroom—but has an urgency we find hard to ignore. A researcher says that this particular sound is a natural one, but "we think that cats learn to dramatically exaggerate it when it proves effective in generating a response from humans."
- The parasite toxoplasma, when it infects a rat, overwhelms its fear of cats and makes it actually feel attracted to them, with predictable results for the rodent. (It then infects the cat, which is where it really wants to live.) Humans infected with toxo also show mental changes: They have an increased risk of traffic accidents, and there's a correlation with schizophrenia. But the most frightening possibility: If it makes rats attracted to cats, does it do the same to people? As one researcher observes of people infected with toxo, "We have a parasite in our brain that is trying to get transmitted to a cat. This changes an individual's personality."

In fact, if you dream of a society where everyone is helpful and kind to others, biologists have some bad news for you: That may be an impossible dream. One researcher, who studied how low-ranking capuchin monkeys use alarm calls to chase high-ranking bullies away from tasty bananas, says we shouldn't be surprised by this sort of thing. It's exactly what's expected, based on the theory that creatures get smarter because group living is all about competition. That idea is called the Machiavellian intelligence hypothesis, and says the capuchin scientist: "One of the predictions of the hypothesis is that deception should be a common behavior."

HEY, BABY, IT'S COLD AND DANGEROUS OUTSIDE

Someone once wrote a book called *Dogs Never Lie About Love.* That's the sort of thing people like to believe. But think about it: The two most basic requirements for survival of a species are food and sex. So if animals lie about food, of course they lie about love as well. It's too important not to.

Some animals have figured out that since an alarm call works to shoo your neighbors away from an attractive food source, it'll also work to keep them away from an attractive lady. Swallows that have read Chapter 3 of this book or who have otherwise learned that their wives can't be trusted may make alarm calls when there's no predator around as a way to frighten off any males from making advances on their mate.

An even more underhanded strategy is that of the male topi antelope. When his date shows signs of losing interest, he'll look in the direction her eyes are wandering, prick up his ears, and snort, using the sound that means he's caught sight of a lion or other predator. A scientist who caught them in the act thought the ruse was pretty lame:

> It made me laugh. It's such an obvious lie—clearly there's no lion.

Obvious, maybe, but apparently it works. When they played recordings, the researchers discovered that females couldn't tell the difference between lying snorts and truthful ones. A snorting male would get two or three more chances at mating, and the guys weren't reluctant to milk it—their snorts were lies nine times more often than they were true.

Alarm calls aren't the only ways animals tell lovers tales. A rooster will make the call that means he's found food when it's really some other object, to attract females to come closer to him. He's careful to make sure she's far enough away that she can't tell that's only a peanut shell on the ground.

And some truly despicable animals do it the other way around, lying about love to get food. Ever watched fireflies on a summer night and wondered what was the point of that magical display? It's firefly singles night. Males fly around announcing their availability, a female perched in the grass responds with the particular flash of their species, and he flies into her arms and together they make little baby fireflies.

Most of the time, that is. Some females will instead mimic the flash pattern of a different species of firefly, one that they can't mate with. Scientists call these "femme fatales," because when the hopeful male responds to her counterfeit invitation, he doesn't get lucky, he gets to be her meal.

SINS OF OMISSION

Animals can lie by not communicating as well. Low-ranking vervet monkeys who see a predator make alarm calls much less often than high-ranking ones do; like the ill-paid staff of bullying bosses, maybe they figure they get treated the same no matter how much effort they make. The monkeys play favorites, too—males are more likely to call to warn a female than another male. And there's also alarm-call nepotism: Females are more likely to warn their own offspring of a threat than other kids of the same age. Ground squirrels, likewise, also often don't bother to warn neighbors of danger if there aren't any relatives nearby.

And by the way, even animals themselves know that this sort of thing is wrong. Some rhesus monkeys don't make the "come to dinner" call when they find food, trying to keep it all to themselves. But if they're caught, they're punished for it: Their troop members beat them up.

APE VS. APE

I know what you're thinking: All of this behavior could still be instinctive. Some ancestral antelope with bad vision accidentally made the lion alarm call when it saw a blurry tan shrub, and he got to leave more of his genes behind by mating with the lady who was afraid to wander away.

But that won't help explain away the orangutans who play tricks on their keepers: One hid an orange in a hand, pretending she hadn't gotten her share. However, once she got a second one, the keeper caught on, and the ruse stopped working. The next day, another orangutan instead hid his orange under a foot so he could show the keeper his innocently empty hands.

In fact, apes that try to fool humans don't seem to have much respect for our allegedly superior intelligence. Long-time chimp observer Frans de Waal tells the story:

> When unfamiliar people are allowed near the sleeping quarters, Jimmie always tries the same dirty trick to lure them. She pokes a blade of straw though the bars and looks up at the stranger with a perfect poker face. The stranger takes the straw, thinking that this is a friendly gesture. At that moment Jimmie's other hand flashes through the bars and grabs hold of her victim. Then the only way to loosen her grip is with someone else's help.
>
> De Waal says they'd never try this on another ape.
>
> "They know each other too well to get away with it. Holding out a straw with a sweet face is such a cheap trick, only a naïve human would fall for it."

While chimps like to play pranks, lazy orangutans lie to get people to do their work for them. A recent study announced the breakthrough that orangutans use pantomime to communicate—and revealed that another way to communicate gives them another way to fool us. One orang pretended it couldn't get termites out of a nest with a stick, and another that it couldn't open a coconut, using gestures to dramatize their sad helplessness. You can't really blame them, since the strategy seems to work: After halfheartedly hitting the coconut with a stick, the second ape handed it to a human and mimed a familiar gesture, and the helpful human opened it for her with a machete.

LET SNEAKING DOGS LIE

Despite these obvious examples, scientists have a hard time proving animals are deliberately trying to trick us, but they're working on it.

One study showed that macaques know exactly what they're doing when sneaking food. Researchers put grapes in two identical clear containers, one with bells that jingled when it was opened, the other, where the bells had the clappers removed. When a nearby person was watching, the monkeys would try to get the treats from either container equally often, but when the observer wasn't looking, they'd take the grapes only from the silent one to avoid attracting attention.

Dog owners may be interested to know that the same experiment done with dogs got the same result. Of course, this probably comes as no surprise to most of us, the lead author of the study included. When her pug steals something, he'll run to another room if chewing it would make noise, but stay put if it's soft and quiet to gnaw. As one dog-owning reader commented on this research, "tell us something we don't already know!"

TECHNOLOGICAL ADVANCES

Perhaps there is one insurmountable boundary for animal liars: They'll never be able to type well enough to pretend they're young and skinny on an Internet dating site. But that doesn't mean they can't use technology to bend the truth.

Some background may be in order, because if you still believe that humans are the only tool-using animal, I'm afraid that's yet another way your knowledge of technology is behind the times. Animals not only use tools, they even make them, and with an impressive attention to detail.

Chimps, for instance, gather tasty termites using a sharp stick to poke a hole in the nest and a frayed one to collect the insects. They're particular about their materials, preferring wood from two different species of tree to make the two different tools. Chimps also make spears to kill adorable little bush babies for meat—"I was flabbergasted," said the touchingly naive researcher who observed this—and they use cooking implements, described as stone and wooden cleavers and stone anvils, to chop food into bite-size pieces.

It seems like it's only a matter of time before we find chimps cooking s'mores over a campfire, but they're not the only primates that use tools to prepare their food. The capuchin monkey uses rocks as hammer and anvil to crack open nuts. They're fussy about their tools too, using the same particularly good sandstone slabs repeatedly, and testing the hammer rocks before use. They'll pick up a stone and heft it, and if it doesn't seem right, toss it away and look for another.

It's not just our close relatives who share these skills, either: There's a species of crow that might be even better than chimps at tool making. One New Caledonian crow became famous when scientists found she would make hooks out of straight pieces of wire to get at unreachable food. In the wild, these

INVERTEBRATE INTRIGUE

Primates may dominate the tool-use stories in this chapter, but it's not only the animals most closely related to us that can use technology to make trouble. Eight flexible legs are almost as useful as opposable thumbs:

- Staff at the Santa Monica aquarium came to work one day to find the place flooded with a couple hundred gallons of seawater. No, not an equipment failure: Their octopus had disassembled a valve, and "grabbed the tube that pulls out the water and caused it to spray outside the tank," said an employee.
- Another octopus caused trouble not just for humans but for his fellow aquarium residents as well. Otto the octopus in Germany was known to juggle his hermit crab roommates and damage the glass of his tank by throwing stones. However, when the aquarium suffered mysterious recurring blackouts, no one suspected a connection. Then Otto was caught climbing onto the rim of his tank and squirting a jet of water at an overhead lamp, short-circuiting it and causing the power to shut down—along with the life-support systems for all his fellow creatures. The light was moved, but Otto constantly demanded new toys to keep him out of trouble—a chess board entertained him for a while, said the director, "But then, he was like, no, I don't want the chess board. And he just threw it out of the aquarium."
- Occasionally these mischievous cephalopods are too clever for their own good. An octopus in San Pedro, California, managed to remove a plastic drainpipe glued in place with silicon sealant. She was found dead in the morning, all the water drained out of her tank.

same crows create devices out of twigs and leaves to probe holes and crevices for insects:

> Crows snip into the leaf edges and then tear out neat strips of vegetation with which they can probe insect-harboring crevices. These tools have been observed to come in three types: narrow strips, wide strips and multi-stepped strips—which are wide at one end and, via a manufacturing process that involves stepwise snips and tears, become narrow at the opposite end.

In fact, even much more simpleminded animals can learn to use tools, if you've got nothing better to do than take the time to teach them. Japanese researchers have trained degus, which are cute little Chilean rodents, to reach under a fence with a tiny rake to get a sunflower seed. They claim that "after learning the basic skill of simply pulling the tool, the degus spontaneously devised more flexible, efficient and versatile use of the tool," moving it in an "elegant trajectory." (We probably don't have to worry about degus taking over our gardening jobs, though: Not only are they tiny, it took about twenty-five hundred trials over two months to train them.)

TOOLS FOR TROUBLE

So if animals can use tools, of course they use them to behave badly. Our cousins the orangutans (who we've already noted making objects to use for sexual pleasure) are perhaps the champions of this. One expert on animal tool use puts it this way:

> If you happen to forget a screwdriver in the gorilla cage, the animals will hesitantly approach it, briefly sniff it, and subsequently ignore it.
>
> Leave it in a chimp cage, and it will be used in vigorous display, thrown about, and forgotten.
>
> But if you leave it in the orangutan cage, one of the animals will unobtrusively pick it up, hide it, and use it to let itself out when you've left for the day.

There's no way to know how often episodes of the latter sort occur; when an orang lets itself out into the keeper area, it's best to try to keep it quiet or heads are going to roll. However some of these apes have used their tool-using abilities to make more public breakouts, such as two that figured out how to avoid getting shocked by the electrified wires around their exhibits:

- At the Audubon zoo in New Orleans, an orang named Berani escaped his exhibit using only a T-shirt. He scaled a ten-foot wall, stretched the shirt out, wrapped it around the hot wires, swung himself over the wires, and climbed the railing.
- At a zoo in Australia, an orangutan disabled the hot wires surrounding her exhibit with a stick, then made a pile of leaf litter and debris and climbed over the wall. Some claimed the escape attempt was due to grief over the loss of her mate, but she'd done the same sort of thing before without the inspiration of tragedy. "We've had issues with her before in normal day-to-day operations where she tries to outsmart the keepers.

She's an ingenious animal," said one curator. And a zoo spokesperson admitted, "She has shorted hot wires before, but just to get food."

Being intelligent animals, both orangs realized quickly that leaving climate-controlled homes that served free meals wasn't such a great idea after all. Karta, the Australian, "sat on top of the fence for about 30 minutes before apparently changing her mind about the escape and climbing back into the enclosure." Berani hung out for about ten minutes, says a zoo spokesperson: "He wanted to explore a little bit and figured it was time to get back home because his zookeeper was yelling at him."

Perhaps these apes have learned their lesson, but the Louisiana zoo decided for safety's sake to change their choice of enrichment materials, realizing that screwdrivers aren't the only things one needs to avoid leaving in the orang cage: "We gave them T-shirts every day," the zoo spokesperson said. "Not anymore."

PLANNING FOR THE FUTURE

Most of this primate badness so far seems fairly spontaneous. This is significant because planning for the future is another of those things we used to think divided humans from other animals. Yet at least as far as primates go, we were wrong about that one too. Combine the ability to plan ahead with the ability to use tools, and you get Santini, a chimp at a zoo in Sweden. He'd occasionally thrown stones at visitors, and why not? As the researchers dryly note, "Stone throwing toward a crowd of people has an instant and dramatic effect."

Anyone who's worked at a zoo full of annoying tourists can empathize with Santini's urge to pitch a projectile, but at one point, the attacks increased to such an extent that staff had to take action. Searching the exhibit, they found caches of stones along the shore of a pond facing the public area. Observing the chimp in secret, they found that before the zoo opened, when no one was watching, he'd collect these stones to use later in the day.

Santini also used the primate tool-making ability to add to his store of weapons: He'd knock on concrete parts of the exhibit till he detected the hollow sound that meant the concrete was damaged by the cold, then hit harder to break off a piece. Sometimes he'd break these chunks into smaller pieces to make more suitable projectiles. At the time, the scientist who reported the situation said:

> It's very hard to stop him because he can always find new stones, and if he can't find them he manufactures them. It's an ongoing cold war.

Later the zoo did come up with a new strategy. Hoping to calm the ape, they decided to castrate him.

NOT THE APES YOU'RE LOOKING FOR

Though many of the stories in this chapter are from zoos, captive apes don't learn how to lie from observing us. They do it in nature too. For example, in the wilds of Borneo, orangutans lie about their appearance at a distance, even without the Internet to help. They use leaves held up to their mouths to make their calls deeper, which makes them sound to predators like much bigger animals. The strategy works because interactions in the dense rain forest often aren't face to face.

In fact, orangutans seem to think humans are even easier to fool about who they really are—they'll try it right in front of our faces. One zoo orang that escaped from its cage into the service area grabbed a floor squeegee when discovered, apparently figuring the keepers would believe he was nothing but a big, orange, hairy employee cleaning up at the end of his shift.

PSYCH OUT

To make a lie work, you need to understand the psychology of your fellow creatures, and orangutans are masters of this as well. One orang managed to make a device to open his cage door latch using just a piece of cardboard. However, the most unnerving part of the story is that he knew enough to hide his tool-making project from the keepers, but realized it wasn't worth the bother to keep it a secret from lowly volunteers.

One of the most accomplished planning, tool-making apes ever recorded turned out to be a master of psychological manipulation as well. Fu Manchu made several escapes a few decades ago at the Omaha Zoo. Positive they'd locked all the doors—and on the verge of being fired for carelessness—keepers set up a secret watch. They saw him pull a door back from its frame, then take out a piece of wire hidden in his cheek and use it to trip the latch. He'd bent the wire into a comfortable shape to hide in his mouth, and had been carrying it around like a set of keys to use whenever he felt like getting out on a nice day.

The most fiendish part of his plot was revealed when the humans tried to figure out where he'd gotten the wire in the first place. They discovered that it came from a light fixture in an adjoining cage, home to an orang called Heavy Lamar. She was housed alone to keep her on a strict diet, but they'd seen Fu Manchu passing her forbidden extra pieces of chow through the bars. Did he offer a straightforward trade, or did he get her used to the treats and then withhold them, taking advantage of her weakness? That part we'll never know.

SEVEN

Masters of Misdirection

ANIMALS WORK TO DIVERT OUR ATTENTION FROM THEIR BAD behavior in many ways. Our species is easily distracted by their grace, majesty, or apparent cleverness. And as many animals have discovered, one of the best strategies is to be cute.

Consider, for example, the rabbit, whose disarming cuddliness distracts us from its other most deadly weapon: Its legendary prowess at producing more adorable little rabbits.

As one reporter on the rabbit-invasion scene insightfully observed, "As a destructive force cloaked in cuteness, bunnies are hard to beat." And no one knows this better than Australians. Domestic rabbits arrived in Australia along with Europeans, and like any good immigrant they seized the opportunities presented by a new land. A dozen or so were released in 1859 and in a few decades they'd spread to most of the country. A seventeen-hundred-kilometer-long rabbit-proof fence built in the first decade of the twentieth century didn't stop them, and it's estimated that by the 1920s, their population had grown to ten billion.

Rabbits ate the heck out of native plants—with the assistance of feral goats they reduced one island to bedrock—and outcompeted indigenous wildlife for resources. While it's hard to keep precise score, they've probably contributed to the fate of many of the twenty-two species of Australian mammals known to have gone extinct as of 2007. At that date, naturalists calculated that of the native creatures that were still hanging on, at least seventeen bird species, thirteen mammals, four reptiles, and 121 native plant species were threatened by the cuddly little invaders.

You'd think we'd have learned a lesson from the foolishness of our ancestors and now know better than to deliberately release nonnative species, no matter how adorable. Unfortunately not. Ask the folks of Long Beach, California, where hundreds of bunnies descended from former pets live on the city college campus. They hop around charmingly and, according to one report, "fight bloody turf wars, burrow deep holes in the lawns, and devour thousands of dollars of landscaping."

These more recent bunny colonizers are helped along by both their cuteness and their reproductive prowess. In Long Beach, one activist experienced the effects of the latter:

> A few years ago, Ms. Olson rounded up 100 rabbits on the south side of campus, and found other homes for them. Only two rabbits on that part of campus evaded her. "Unfortu-

nately, one was male and one was female," she said. Within six months, the population on that side of campus had climbed back to 100 rabbits, she said.

North of the border, the University of Victoria in British Columbia had a similar problem, where an estimated sixteen hundred rabbits were destroying a prized rhododendron garden. When officials tried to reduce the population, they were blocked by the mesmerizing effects of cute: Bunny-hugging humans actually got a legal injunction that temporarily delayed trapping.

BEAUTY IS ONLY FEATHER DEEP

Cuteness isn't the only quality that animals use to blind us to their nature; we're also suckers for natural beauty. And what's more beautiful than a hummingbird? To their fans, these tiny, colorful creatures are as magical as fairies. Wildlife writer Richard Conniff met a birding guide who knew these deluded souls well:

> "These creatures have a following like mythical beasts," said one of the guides, a little ruefully. "There are people who don't care anything about birds, or other wildlife or nature, but they love hummingbirds. We had one woman tell us: 'I just love hummingbirds and unicorns.' And I don't think she drew any distinction between the two."

However, the birding guide harbored no such illusions: "We're probably lucky these things aren't the size of ravens, or it would not be safe to walk in the woods." As we saw in Chapter 3, hummingbirds practice prostitution and will knock a female out of the air to rape her; they also use their beaks, claws, and long pointed bills as weapons against rivals. This sort of thing is no secret to science, as Conniff discovered:

> A scientific paper about the rufous hummingbird includes this endearing notation: "SOCIAL BEHAVIOR: None. Individual survival seems only concern."

What makes hummingbird fandom all the more mysterious is that you don't need an advanced degree or a foreign expedition to observe the creature's true nature.

One website written by hummingbird aficionados describes them puffing themselves up to look larger, using their beaks and claws as weapons, and being so aggressive toward their own kind that they attacked a plaque decorated with fake hummingbirds. If you want to see for yourself, they advise, set out a large number of feeders, spaced in a way that makes it impossible for one bird to guard them all, and step back:

> Don't bother to try and stop them from fighting. It's best to just leave them alone and let them work it out. We have a rule on the top deck of the World of Hummingbirds: Hummingbird Farm of "no body-slamming." Whenever the hummingbirds start to body-slam each other, we yell, "Hey." Now they just do it when they think no one is looking.

MENDACIOUS MAGNIFICENCE

Cuteness and magical beauty aren't the only qualities animals use to cloak their true nature: Some go for power and majesty instead. One example is the noble lion, which as we saw in Chapter 4, has the charming habit of killing baby lions. The king of beasts has other flaws as well, including extreme laziness. *Smithsonian Magazine* profiled a scientist who has spent his life studying lions and whose very first research conclusion was rather disappointing:

> When he first visited the Serengeti lions in 1974, he concluded that "lions were really boring." The laziest of all the cats, they were usually collapsed in a stupor, as if they had just run a marathon, when in reality they hadn't moved a muscle in 12 hours.

And if you haven't already had enough of your illusions smashed about these supposedly brave beasts, here's what one journalist discovered about what they're afraid of:

> Earlier I had asked what kind of anti-lion gear the researchers carried. "An umbrella," Jansson said. Apparently, lions don't like umbrellas, particularly if they're painted with large pairs of eyes.

Fearsome reputations with no foundation go way back in evolutionary history. It turns out that even impressively frightening extinct animals don't live up to their press. Picture the *Tyrannosaurus rex*: mighty carnivorous giant, battling fierce rivals to the death! Or . . . not so much. A couple of scientists looked at the fossilized contents of *T. rex* stomachs and droppings, then put those data together with the rarity of bite marks on the bones of large prey species and the surprising absence of fossilized young given the large numbers of eggs that dinosaurs laid. Their conclusion: *T. rex* must have been another of those species with a spectacular press agent, because in reality, you know what they killed and ate? *Little tiny defenseless baby dinosaurs*, of course.

FAMILY SECRETS

It's a fact of life: We're often the last to know the secrets of those who are closest to us. Perhaps this is why we've been so blind to the faults of our primate cousins. Yet despite repeated revelations about the true nature of our fellow apes, we can't seem to get out of the habit of putting them on a primate pedestal.

For a long time, scientists used to believe that humans were uniquely violent. It was conventional wisdom that other animals, even those that share most of our DNA, like chimps, killed for food or competed for mates but they didn't assault or murder their own kind just for the heck of it and certainly not for control of land. Then in 1974, an observer at Jane Goodall's research station in Tanzania saw eight chimps cross into another troop's territory, come upon a lone male, and attack him. One of the intruders held him facedown in the dirt while the others bit and pummeled him for a good ten minutes.

IF THEY DO THE CRIME . . .

Some academics have argued that dolphins are so intelligent that we need to treat them as more than animals, based on evidence that they have large brains, pass on cultural activities, recognize themselves in mirrors, have sophisticated communication systems, and so on.

Setting aside the implication that it is okay to treat someone badly as long as they're stupid—because honestly, who hasn't had the temptation—I am happy to grant the premise that dolphins are brilliant, since it takes nothing away from my argument that they are basically bastards. In fact, I'm quite intrigued by the suggestion of one author that these evil geniuses should be considered "non-human persons."

"Dolphins should be considered non-human persons," says ethicist Thomas White, "because they have the kind of consciousness that, in the past, we thought was unique to our species. They're not just aware of the world around them but they have the ability to look inside and say 'I.' They have a sense of choice and will."

White argues that dolphins, like humans, should be given "moral standing" as individuals, and surely any reasonable person who has read this chapter would support this. Yes, we've seen that dolphins are gang rapists, murderers, and baby killers and pose a serious threat to innocent, ignorant humans—but that's exactly the point. It's about time we start treating dolphin persons in the same way that we treat human persons who commit those crimes. Enough putting them on T-shirts, then, and more putting them on trial: Let them pay the price like the rest of us have to.

It turns out that this kind of raid was no isolated incident, but rather a regular feature of chimp territorial warfare. Another researcher witnessed eighteen chimp murders committed in the course of ten years and found that the attacking group would then take over the home turf of the victim, eventually increasing their troop's territory by over 20 percent. And how about the bonobo, the chimp with the allegedly blissful, un-hung-up sex life? They supposedly resolve conflicts without violence by making love in all possible combinations as casually as we shake hands in greeting. We thought they treated other primates equally peacefully, grooming monkeys and playing with their babies—but now they've also been seen hunting monkeys in a pack to kill and eat them.

At least one expert, a primatologist at the Lincoln Park Zoo, who heard about this discovery, had the sense to suspect these new-age primates of harboring dark secrets all along: "The second I read this, I thought: Oh good, finally! I was just waiting for something like this to come up."

And another sexually enlightened species of primate has also turned out to be a disappointment: Scientists thought they'd found a monkey in Brazil with such a peaceful, egalitarian society that the males waited patiently in line to mate with females. They even referred to the muriqui as the "hippie monkey."

They just hadn't been looking long enough. Sure, these monkeys were peaceful when there was enough of everything to go around—who isn't? But at another study site, mates were harder to come by, with no ladies offering their favors to multiple males in a row. There, researchers saw behavior just like that of chimps, including an unprovoked gang attack on a lone individual who was bitten and pummeled to death. The unlucky fellow was attacked over his whole body but particularly, it seemed, his genitals. That's what happens when you deprive the hippies of their free love.

DECEIVERS OF THE DEEP

The more we love and admire an animal, the worse the truth is likely to be. Two species in particular have done such an astonishing job of manipulating our minds that they deserve individual attention. One has even managed to worm its way into our very homes, and for this accomplishment the dog gets the whole next chapter to itself. For now, though, we will take some time to examine another—an animal we have raised almost to the level of a mythological creature, ignoring the dark secrets it hides behind its smiling exterior: the dolphin.

"Everybody who's done research in the field is tired of dolphin lovers who believe these creatures are floating hobbits," said renowned animal trainer Karen Pryor, quoted in the *New York Times*—in 1992. She must be a whole lot more tired now, because nothing has changed.

WHAT'S LOVE GOT TO DO WITH IT?

Consider the courtship behavior, if you can call it that, of these supposedly gentle creatures. For obvious reasons it's difficult for humans to observe dolphin sex, so some of the details still remain obscure. But given what we do know, that's probably just as well. You may read descriptions of dolphins "nuzzling" and "caressing" each other, but it's pretty clear that not all dolphin sex is so tender—or even consensual.

Ever been impressed by dolphins leaping and somersaulting in unison at an aquarium show? They do that in the wild too. First they gang up to capture a female from a rival group of males. Then to keep her where they want her, they perform those aquarium-show acrobats as they circle around, forming a moving perimeter that she can't get past. If that's not enough, they'll chase, bite, and body slam her. "Sometimes the female is obviously trying to escape, and the noises start to sound like they're hurting each other," says one expert. "The hitting sounds really hard, and the female may end up with tooth-rake marks."

We're also impressed by dolphin communication. Wouldn't it be amazing to decipher it and learn what these remarkable creatures are saying? Well, you might be sorry you asked when it comes to their conversation on a date:

> Sometimes a male will make a distinctive popping noise at the female, a vocalization that sounds like a fist rapping on hollow wood. The noise seems to indicate "Get over here!" because if the female ignores the pop, the male will threaten or attack her.

And apparently it's also fairly obvious what the female is trying to say in return. In one of the few cases observed close-up, a dolphin communication researcher described a female's vocal reaction to what looked like a rape attempt:

> She would basically put her genitals above the water and she'd keep putting her belly above the water so the males couldn't get to her. And as she did that there was a really intense emission of whistles, and basically it was the same whistle being called over and over and over again and it was quite a distressed type of call.

BUOYANT BULLIES

If we ever do learn to translate dolphin language, we're sure to find that they have no words for "pick on someone your own size." Scottish researchers who found stranded dead bodies of harbor porpoises first considered boats and fishing nets as the likely causes of their injuries. As one researcher describes it, these small marine mammals were "suffering massive haemorrhages and their internal organs were all mushed up. They were being battered to death, really."

Then a corpse turned up that also had recognizable bite marks, perfectly matching the spacing of bottlenose dolphin teeth. Since then, attacks by dolphins on their smaller relatives have actually been witnessed, and postmortem exams in one part of Scotland showed that 60 percent of the dead porpoises found had been killed by dolphins.

The researcher just quoted is conducting acoustic research to try to figure out why dolphins commit these brutal murders. My hypothesis would be "because they're assholes," but a more

DOING THEIR EVIL BIDDING

Dolphins are not only bad to humans and to other animals, but in perhaps their most fiendishly clever accomplishment, they have even convinced us to be bad to other animals on their behalf.

The successful campaign for dolphin-safe tuna was based on outrage that this intelligent creature was being caught in nets used for tuna fishing and killed. Now, most fishermen no longer locate tuna schools by looking for the dolphins that follow them, nor do they encircle the lot of them with a huge net and drown dolphins by the thousands for the sake of your lunchtime sandwich.

Problem solved, right?

Not so fast. So, how do fishermen find those tuna now, instead?

The ocean is so huge that animals living there can go their whole lives without encountering a solid floating object. If you put one down, they have a tendency to come and check it out en masse. Who can blame them? They've never seen anything like it.

So that's the new trick. The industry uses "fish-aggregating devices" that get them all to gather around. However, tuna aren't the only ones that show up to the party; the objects are equally fascinating to endangered sea turtles, at-risk shark species, rays, and other kinds of fish—as well as small tuna that haven't lived long enough to reproduce yet. With the new method of fishing, for each dolphin that's been saved, here's what's estimated is killed instead:

- 25,824 small tuna
- 382 mahi-mahi
- 188 wahoo
- 82 yellowtail and other large fish
- 27 sharks and rays

- 1 billfish
- 1,193 triggerfish and other small fish
- 0.06 sea turtles

Oh, and those dolphins that were being killed when we did it the old way? The species were mostly spinner dolphins and spotted dolphins, which are *not even endangered*. Hey, sharks and rays and mahi don't have that hypnotic smile, so what's a primate to do?

specific reason has also been suggested. Scientists who examined the bodies of dead dolphin calves found that their injuries didn't look like they were caused by predators, boats, surf injury, or similar causes. Rather, they looked exactly like the injuries previously found on harbor porpoises.

So it's been suggested that dolphins kill the harbor porpoises for practice, so they'll be good at killing their own babies. Which they probably do for the same reasons as many other animals, as we saw in Chapter 4. Dolphins breed infrequently, and who wants to put the effort into raising another guy's offspring? As one researcher notes, if the female you've got your eye on already has a kid, "there's no point in hanging around for two years with the wrong male's calf."

PERVERTS WITH FINS

Dolphin expert Karen Pryor has observed, "the sentimental view that these animals are harmless stems at least in part from the fact that they are usually in the water and we are usually on boats or dry land: they can't get at us." But while most of us have the sense not to walk up to a lion on the savannah, many are deluded enough to purposely get in the water with dolphins, and that's where the problems start. Yes, many swim away from their dolphin encounters unharmed, sharing stories of "unconditional love, peace, and bliss" as they looked into that smiling cetacean face. There are even stories of helpful dolphins saving desperate humans from drowning. However, as more than one rational individual has observed, there's a large gap in the evidence; you're not going to hear from the people who the dolphins push *away* from shore.

However, there have been victims of dolphin hostility who lived to tell the sobering true tale. A dolphin named Georges became a tourist attraction at a harbor in England in the early 2000s, attracting thousands to watch him and swim with him. All very nice and no doubt good for the local economy—until Georges began to show what was delicately described as "an unhealthy interest" in divers.

To put it more bluntly, Georges was attempting to mate with them. An expert who was called in, familiar with the male dolphin courtship techniques described above, warned of the danger: "When dolphins get sexually excited, they try to isolate a swimmer, normally female. They do this by circling around the individual and gradually move them away from the beach, boat or crowd of people."

Bad enough if you're a female dolphin—for a human, a good way to be drowned. Visitors were warned of the risk, but they persisted, listening to people who called their encounters a "miracle" and "most amazing thing I've ever done" instead of to those who were bitten, hospitalized, or saw their children pushed into deep water. As a rescue diver said in exasperation, "It's quite obvious that people are getting hurt and they are getting out of the water but 15 minutes later they are getting back in."

EIGHT

With Friends Like These . . .

IF DOGS ARE MAN'S BEST FRIEND, OUR SPECIES NEEDS TO TAKE a hard look at our self-esteem problem.

Canine bad behavior is so routine that there is a whole genre of self-help literature for coping with it. Millions of words have been written about providing clear consequences and making sure that dogs don't get a free ride. But the advice goes unheeded again and again: One look at that wagging tail and sad eyes, and we pat them on the head and forgive them. Then the minute we turn our backs they're at it again, knocking over the trash, peeing on the carpet, and chewing our possessions to shreds. And as we'll see in this chapter, that stuff is just the start.

Don't we deserve to be with someone who treats us better? Take a hard look at the evidence, and maybe you'll feel empowered to take the first step out of this dysfunctional relationship.

NO ACCIDENT

Most people are positive that their dogs would never bite them. They're probably right. Why do something that obvious when there are much sneakier ways to hurt someone? According to the Centers for Disease Control and Prevention, more than eighty-six thousand emergency room visits per year are due to fall injuries caused by pets. And if your dog tries to convince you that those cats underfoot are to blame, note that dogs were almost 7.5 times as likely to cause injuries as cats—88 percent of the emergency room visits studied were the fault of a dog.

What's especially sinister is that these incidents seem totally innocent. Biting may result in unpleasant consequences for the biter, but people don't blame the dog when they trip over it. One woman broke her ankle when chasing her puppy around the dining room and spent seven weeks in a cast, but—while still walking with a cane and undergoing physical therapy—denied that the dog was responsible:

> I can't blame the dog. I can only blame myself and the slippery floors . . . I was angry at myself for trying to keep pace with a 10-month-old puppy as a 44-year-old woman.

Still, sometimes living with us is so aggravating that dogs are going to lose it and use their teeth. And we ought to admit that we knew this going into the relationship. As one evolutionary biologist said, "It's difficult to domesticate a large carnivore; they tend to eat your children and bite you." Yet if we thought that selective breeding for small and cute would solve the problem, it didn't exactly work out that way. While uninformed lawmakers think that pit bulls need to be banned, one careful study found that the breeds with the most aggression toward humans were Chihuahuas, dachshunds, and Jack Russell terriers. All cute little pups you can carry in your arms—if you're brave enough.

DOMESTIC VIOLENCE

When dogs do bite, they're generally just trying to keep us in line, not eliminate us entirely—we're much too useful to them. Despite enormous media attention devoted to fatal dog attacks, you're five times more likely to be killed by lightning than by a dog, no doubt because lightning doesn't need us to open its canned food.

Still, it's fair to point out that even in Australia, where it seems like all the wildlife is specifically designed to kill you, dogs are statistically the most dangerous animal. The continent is home to sharks, crocs, venomous snakes, spiders, and jellyfish and less obvious terrors as well: Even the cute platypus has venomous spines on its feet, and someone was recently attacked so severely by a wombat that he had to fight the animal to the death with an ax.

But the numbers show that even Down Under, you're much more likely to be injured by your dog than by any of those hazardous beasts. There's an average of one shark fatality a year, but between 1997 and 2003, eleven people were killed by dogs. And even when it comes to lesser injuries, dogs stand out: On average, 3,867 hospital visits each year are due to stings and bites by snakes, spiders, insects, and plants combined, but dogs by themselves are to blame for 2,000.

And the worst part? In the majority of cases the dogs go for friends and family. At least the wildlife is attacking strangers.

GUNS DON'T KILL PEOPLE . . .

Anyone who has a dog knows that it understands many of the advantages of human technology. From rides in the car to tennis balls to food that you don't have to run down and kill, civilization has been very good to canines. And some dogs also recognize that human weapons are an advance over natural ones. Unlike teeth, guns have long-range capabilities, and of course, no one believes you did it on purpose:

- In New Zealand, a man on a hunting trip was shot in the butt by his own shotgun when his dog jumped into his truck after him and stepped on the trigger.
- A teenager lost several toes when the same thing happened to him on a hunting trip in Arkansas.
- In Oregon, a man on a duck-hunting trip was shot in the legs and buttocks when his dog jumped into his boat. The shotgun blast blew a hole in the boat, too.

And because authorities can't believe that a dog could be to blame, suspicion often falls on someone else:

- A woman in Florida said that her husband was accidentally shot and killed when the family dog knocked over a rifle. She was charged with murder.

A HUNGER FOR LEARNING

Do dogs really eat homework? Students who use this excuse are inevitably considered unreliable witnesses. But others whose motives are less questionable have also told tales of dogs consuming irreplaceable paperwork. A dog ate the only copy of the first draft of John Steinbeck's book *Of Mice and Men*, although typically, the author made excuses:

> *I was pretty mad, but the poor little fellow may have been acting critically. I didn't want to ruin a good dog for a manuscript I'm not sure is good at all.*

And a man in South Carolina was unable to run for a position on the board of education when his puppy ate the petition with more than two hundred signatures on it that was needed to validate his candidacy—the night before it was due.

Other sorts of valuables eaten by dogs are sometimes recoverable, if the victim is willing to suffer through the process. When a North Carolina woman who'd lost $400 cash found parts of the bills in her dog's droppings, she began to collect and wash them in hopes of having enough pieces to exchange for new bills.

It's a little easier for those whose dogs eat less digestible assets, but still, the owner of a dog who swallowed a $20,000 diamond in a Maryland jewelry store reported mixed feelings. "It was not that pleasant. I followed him; I had to pick up his stuff; I had to go through the things," he said, but victory was sweet, if stinky: "I can understand what it was like in the old Gold Rush. I felt like I had just hit pay dirt."

- A man in California was charged with involuntary manslaughter in the death of his wife, despite his claim that his dog tripped him while he was holding a pistol.

In the latter case, the trial ended in a hung jury, with three members voting to convict but nine others—no doubt dog owners themselves—voting not guilty.

"WORKING" DOGS

Even when dogs seem to be obeying our orders, they're not always doing quite what we think. Take dogs that guard flocks of sheep and other livestock. Experts studying these canine employees find that they do get the job done, but it's not because they work hard at presenting a courageous and fearsome front:

> Livestock guarding has little to do with the legendary brave companion fiercely protecting its master's property. Rather, guarding dogs protect by disrupting predators by means of behavior that is ambiguous or contextually inappropriate: barking, tail-wagging, social greeting, play behavior, and occasionally, aggression.

Translation: These dogs aren't guarding, they're just goofing around, and the predator's reaction is less fear than it is, "WTF?"

Unfortunately, there's little hope of getting people to recognize such subtle subversive behavior, since some don't see the problem even when it bites them on a limb. A policewoman in England responded to an armed robbery at a pub with her trusty police dog, and when she was shot in the leg by one of the thieves, she commanded him to attack. How was her confidence in her canine colleague rewarded? He turned on her and bit her arm as the pair of thieves escaped. The officer was sure it was all an innocent mistake—she said of her dog, whose name should perhaps have been a warning: "I don't blame Chaos at all."

CANINES AND CARS

It's easy to shut our eyes to the psychological harm caused by dogs. But it's much harder to be in denial about the joy our friends take in wrecking our personal possessions, since the evidence is so tangible. Still, shredded couches and chewed-up shoes may not prepare your mind for the frequency of one particular sort of canine property damage. As we saw in Chapter 2, many animals interfere with traffic—but only one does it by actually doing the driving:

- In New Zealand, a man left his dog in his vehicle with the motor running while he ran into a store to buy beer. His dog shifted out of park and drove through the doors of a neighboring café.
- In Oklahoma, a dog waiting in a car at a car wash shifted it into reverse, backed it out of the car wash bay onto the highway, then looped around, still driving backward, back to the car wash. (The owner's relief at the miraculous lack of damage was short-lived: Her vehicle was impounded by responding officers when she was unable to provide proof of insurance.)

THE MENACE BEHIND THE PURR

The eternal war between dogs and cats is nothing next to the never-ending debate between their owners, and this chapter should in no way be taken as supporting evidence for feline superiority. In fact, even when it comes to traditional canine bad behavior specialties, cats actually give dogs quite a good run for the money. The most iconic example of this is the attack on the mailman.

In England, these incidents frequently make the news when the household of a repeat offender has its mail delivery cut off. And if you've read this far, you probably won't be surprised to learn that the cats are only part of the problem—most of the owners deny any culpability on the part of the beloved kitties, reacting with either disbelief or, yes, laughter:

- In August 2002, a letter carrier attacked by cats called Boo Boo and Yogi told a dramatic story:

  I put up with it the first couple of times, but the last scratch was quite deep. Blood was dripping on to the driveway and over other letters in my bag. After the attack the cat jumped up on the window sill and looked out at me as if to say "got you that time."

  Their owner was incredulous:

  Mr. Davies said: "I can't understand the attacks. They are both really well behaved cats but are very playful."

- In June 2009, another owner pooh-poohed a letter threatening suspension of delivery: "We were not around when this happened, but it seems some mail was put through the letterbox and their hand was scratched. Illy is only a kitten and I am sure she was just playing. . . . Everyone finds it so amusing that our playful kitten has been mistaken for some savage beast."

- In October 2009, Magic dashed out his cat door after the carrier three days in a row, but his owner was indignant about the suspension notice: "I told them my cat wouldn't hurt a fly. . . . He's a soft cat. We've never seen him attack someone and we've never heard of him hurting anyone before. I can't believe they are saying this."
- In April 2010, Tiger's owner scoffed at reports that he was assaulting mail carriers as they approached the door and chasing them down the path: "Tiger is 19 years old, he dribbles when he sleeps and snores—he sleeps for 20 hours a day."
- In November 2010, another family claimed that their cat had been "playful" when it swiped at the carrier's hand as he was pushing mail through the slot. The mother insisted that the cat Lana, who the children playfully call Lana Banana, was "really docile, I can pick her up like a baby and she won't bat an eyelid."

Rarely, we do hear of attack-cat owners who admit they have a problem. In 2007 the owner of Dipity took the news with resignation when she got the letter threatening to suspend her delivery service. "I can't say I blame them for threatening to cut me off. I love Dipity to bits—she's adorable—but I'd be the first to admit she's a little terror," she said. "All she wants to do is pick fights. When I took her to be neutered she tried to gouge lumps out of the vet and was hissing at all the dogs."

And we can only hope that more will follow the example of the owner of Blackie, a cat who counted at least five postmen, one police officer, five paperboys, one takeaway driver, and one

continues on next page . . .

construction worker as his victims. In December 2006, she finally took responsibility:

> *At the door to Ann Hogben's home in Ramsgate, Kent, is a newly erected sign which simply reads: "Warning: Dangerous Cat—Has Attacked 13 People in the Last Six Years."*
>
> *The turning point came in the last few weeks when Blackie attacked a postman as he pushed envelopes through her letter box. Unaware of what lurked behind the door his latest victim was left with a series of open wounds.*
>
> *"I came home and Blackie was sitting there like butter wouldn't melt in his mouth, but I knew something was wrong because he had blood on his claws."*

The owner insists that with her, Blackie is as cuddly as can be, but she acknowledges that he has "problems with people in uniform" and "authority issues" and said from that point on, in addition to the sign, she'd be locking Blackie in another room whenever there is a knock at the door.

- On Long Island, New York, a man left his car running to keep his dog warm while he went into a coffee shop to sign up for an open mike night. Perhaps as a comment on his owner's talent—or a desperate attempt to keep him from publicly embarrassing himself—the dog put the car in gear and drove it into the store window.

Similar mishaps can occur even if you're careful never to leave the dog alone in your vehicle:

- When an Idaho man arrived home from picking up a pizza, his dog jumped in, knocked his Chevy Impala into gear, and drove it down an incline into a river. The dog, no fool, jumped out along the way and was unharmed.
- In a brilliant combination of two types of bad dog behavior, a man in Minnesota lost control of his car and drove it into a power line pole when his dog began "throwing up all over him," a story the police were careful to corroborate by finding vomit in the car.

And some of these dogs seem to have intentions more sinister than a mere joyride: When a Florida man put his pickup in neutral and got underneath it to look for an oil leak, his bulldog jumped in, put the vehicle in gear, and ran over him, sending him to the hospital.

Don't say you haven't been warned.

ONLY HURTING THEMSELVES . . . AND YOUR WALLET

Dogs have a particularly special talent for another kind of bad behavior: eating everything in sight, whether it's actually food or not. If they were wild animals, of course, this would be their own problem. But since they've opted into this sweet domestication deal, what it means is that when we aren't paying our own medical bills from all the car accidents, falls, and shootings, we're paying the dog's instead.

There's a pet health insurance company that gives the Hambone Award every year to the most outrageous claim, and as the name suggests, many of the nominees are poster pups for inappropriate canine consumption. The 2010 winner, Ellie the Labrador, ate a beehive full of dead bees after exterminators had sprayed it with pesticide. The vet said she'd be fine, but the owners were the ones who had to deal with the aftereffects:

> She acted just fine that week, really, but every time she went to the bathroom, she pooped bees. Thousands of bees. I don't know where they all came from—the hive wasn't that large.

Ellie beat out a number of worthy competitors including:

- A terrier that bit a chain saw—while it was in use
- A Labrador that ate twenty-three packages of instant breakfast mix
- A poodle that ate two baby bottles and a dirty diaper, whose owner said:

  > We thought we were being careful by putting the bottles underneath a metal cover in the sink and using a separate trash can for the diapers, but we underestimated Roscoe's determination.

Typically, owners will dismiss these incidents as innocent mistakes, but some cases present pretty convincing evidence against that rationalization. The owner of an earlier Hambone Award winner, Lulu the bulldog, assumed that all of those baby pacifiers had gone missing by accident. So when she saw Lulu eat one, she rushed to the vet, only to have the surgeon discover *fifteen* of them in Lulu's stomach. I think we can agree that anyone could make the mistake of eating *one* pacifier. But when you get up into the double digits, there is clearly some deliberation involved.

CANINE BRAINS

One thing we can't really blame dogs for, though, is that they take advantage of our delusions about them. We offer free food and lodging and ask almost nothing in return, and who'd be dumb enough to turn down a deal like that?

However, dogs aren't nearly as smart as we like to think they are. The pioneering psychologist Edward Thorndike observed way back in 1911, "Dogs get lost hundreds of times and no one ever notices it or sends an account to a scientific magazine. But let one find his way from Brooklyn to Yonkers and it immediately becomes a circulating anecdote." There are ways in which dogs are clever, it's true, but the details are rather telling. Experiments where they perform better than other animals tend to have one thing in common—the dogs are using us:

- In a test where there are two containers and only one has food in it, chimps don't seem to get it when a human points to one of the containers. They'll still pick either one to try to get a treat. But dogs know what the pointing finger means almost immediately: We've done the work finding their food, just as they expect us to.

- In another study, both dogs and wolves were presented with a piece of meat in a cage, with a rope attached, its end sticking out. Both kinds of canine quickly learned to get the meat out of the cage by pulling on the rope. Then the experimenters sneakily made the task impossible by fastening the rope in a hidden place. Wolves just kept on trying futilely on their own, but dogs would make a brief attempt and then turn to look at the human. Their expression was no doubt familiar to any dog owner; it was the one that means you're supposed to come over here and do it for them.

It's also worth noting that the dog demanding your help getting the caged meat—or begging for a piece of your steak—is fibbing when he tries to convince you that it's critical to his nutritional needs as a carnivore. In fact, it's been quite convincingly argued that dogs are not mighty meat-eating hunters like their lupine cousins. They're actually scavengers that evolved by hanging around human settlements, domesticating themselves in the process of eating our garbage. The researcher responsible for that theory makes this observation about our dumb, trash-eating, manipulative best friends:

> The wolf has a big brain; the dog's got a little tiny brain. Well, who in the world has little tiny brains? Animals that don't need brains. And the dog, you know, a scavenger, doesn't need much of a brain. I mean, it doesn't take a lot of cunning to figure out where a rotten tomato is.

And yet, dogs shoot us with our own guns, run over us with our own vehicles, and still expect us to cheerfully serve them—and we do it. Ask yourself: Who's the stupider one in this relationship?

NINE

Our Own Worst Enemies

IN MANY CASES, THERE'S A VERY SIMPLE REASON WHY ANIMALS behave badly: Nature sucks. Unfortunately, human beings not only have lost sight of this fact but have managed to turn it completely backward. We think of nature as the unspoiled opposite of civilization. Advertisers even use the word *natural* to make you think "safe" or "gentle" or "healthy"—in short, something that isn't going to hurt you.

But civilization is the opposite of nature for a reason—we invented it because nature is *dangerous*. In nature, something is always trying to kill you, because that's basically how the system works. The cow rips some grass from its comfortable home and turns it into cow, then we slaughter the cow and turn it into hamburger, then the heart attack from the hamburger kills us and the worms crawl in and out and turn our dead body into worm poop that fertilizes the grass. That's the circle of life. In some ways it's beautiful, but the one thing it isn't is safe and cuddly.

If you still feel a resistance to this notion, consider the following:

Cobra venom	Twinkies
Ebola virus	Tupperware
Lightning	Toilet paper
Bears	Teddy bears

Which set is natural? Okay, now: Which set is better at killing you? Exactly. Nature does quite well on her own, thank you.

SELF-SACRIFICE OR SELF-INTEREST?

Our ancestors couldn't get away from nature, and their reaction wasn't gratitude for how easy it was to stop and smell the flowers. Instead, they were inspired to invent buildings, central heating, medicine, guns—all meant to give our species a fighting chance against the totally natural things that were constantly trying to kill us.

It turned out that there are unintended consequences of being comfy and protected from the natural world, though, and one of these is that we've forgotten what its residents are really like. If our distant ancestors, living close to nature, had heard of a dog that stayed with a child lost in the woods on a cold night, they'd never have written these headlines:

PUPPIES SAVE THREE-YEAR-OLD BOY LOST IN FREEZING VIRGINIA WOODS

LOST CHILD DISCOVERED NEARBY WITH LOYAL DOG

BOND WITH DOG HELPED GIRL SURVIVE

Saving a child from freezing to death does seem heroic, until you think about it for about a minute. Because if you've never noticed that your pets are far more cuddly and affectionate in cold weather, you're just not paying attention. (If it's cool in the house right now and there's a dog around, go lie down on the couch and see what happens.)

It's nice that this behavior is helpful, of course, but its motivation is totally self-serving. Dogs in past generations that wandered away from other warm bodies at night and died of hypothermia did not get to pass on their genes. So dogs alive today had great-great-grandparents that curled up next to other warm bodies on a cold night, and naturally they do the same.

What about the much-praised loyalty of those child-saving dogs? Here's a "professional dog trainer" quoted in a story about three-year-old Victoria Bensch of Arizona, who wandered away from home one afternoon and wasn't found till the next morning.

> "This was a deeper connection the dog had with the child," Coddington said. "Otherwise, if that connection wasn't there, it might not have happened the way it did."

The trainer goes on to explain that the dog felt responsible for Victoria as a member of his pack. But then why did a stray dog in the Yukon wilderness do exactly the same thing with a boy he'd never met before who was lost during a family camping trip? It's simple—you don't need a "bond" or "connection" for a dog to appreciate the fact that you radiate a nice warm 98.6 degrees.

ASKING FOR IT

So, since we no longer live in unheated shacks that we share with our livestock, we don't understand the natural behavior of a cold animal anymore. But what's the harm of seeing it through rose-colored glasses? It certainly makes it easier to put up with some of the other things our dogs do. And as for other animals, if our ignorance is based on the fact that we hardly ever encounter them in natural situations, what's the problem with believing that those faraway creatures are kind, noble beings frolicking in a land of rainbows and flowers?

Well, what's wrong is that since we haven't quite managed to drive *all* our fellow creatures to extinction yet, people still do run into them from time to time. And believing that nature is benevolent and animals are noble and cuddly is likely to get you into some serious trouble. In some of the stories of animal assault in previous chapters, victims were at first put off their guard by the cuteness factor. Sure, seeing three otters in the wild is totally cool—till someone ends up in the emergency room getting rabies shots.

Some people don't even wait for the animals to make the first move. A man who took a seven-hour train ride just to see the panda at the Beijing zoo was overcome with a "sudden urge" to jump into its enclosure to pet and hug it. The panda, of course, reacted naturally, and—in another example of how nature is not always gentle, safe, and good for you—bit the man on both legs, putting him in the hospital. The panda-hugger explained his thinking this way: "No one ever said they would bite people. I just wanted to touch it."

Another panda-hug perpetrator at a zoo in southern China said, from his hospital bed:

> Yang Yang was so cute and I just wanted to cuddle him. I didn't expect he would attack.

It was not the first time Yang Yang had bitten someone who got too personal, but a zoo official was emphatic about where to put the blame:

> He said it was not clear whether the facility would add more signs around the enclosure or put more fences up.
>
> "We cannot make it like a prison. We already have signs up warning people not to climb in," he said. "There are no fences along roads but people know not to cross if there are cars. This is basic knowledge."

KEEPING UP WITH INTERNATIONAL COMPETITION

The Chinese are hardly the only ones that are easily blinded to danger by a furry coat, and don't think that leaping into zoo enclosures for a cuddle is an exotic foreign impulse, either. Western zoo patrons have also found inappropriate ways to express their affection for dangerous animals:

- A woman (later found to have a blood alcohol level twice the legal limit) ignored warning signs and barriers and had several fingers bitten off when she tried to feed a bear at a zoo in Wisconsin, no doubt making it a memorable trip for her three-year-old granddaughter. (She was later fined for violating the zoo's rules.)

- A visitor climbed over a fence, scaled an eight-foot-tall rock structure, and bypassed an electric wire to jump into the elephant exhibit at Cameron Park Zoo in Texas, where no one blamed the elephant for smacking her upside the head. "They're not used to somebody being in their space," said the zoo's director.

Other zoo patrons have climbed into the exhibits of snow leopards and lions, offered body parts through fences to jaguars and otters, and one was bitten because he was trying to "help" a loose gibbon at the Cincinnati Zoo.

CODEPENDENT CREATURES

Our delusions about the nature of animals don't always end in bloodshed, fortunately. It's true that some of the offenses are minor and inconvenience only a few. A penguin named Kentucky at a zoo in England has become a favorite because of his troublesome dislike of water:

> "It's a bit too cold for him in the water, so he spends all his time on the rocks just walking around," said Adam Stevenson, the zoo's assistant bird keeper. "It's a bit of a pain having to

> go over especially to him to feed him because he won't go in the water, but he's a real character and everyone at the zoo loves him."
>
> Stevenson said keepers douse Kentucky with water at least twice a day to keep his feathers healthy and clean.

True, this penguin is making extra work only for a handful of zookeepers (which, as a career move, might actually qualify him for a promotion to management). But in many other cases, behavior that is a wider menace to the general public is seen as similarly charming. In New Jersey, a wild turkey took up residence on exit 14B of the New Jersey Turnpike and caused problems by dashing in and out of traffic and prompting people to do the same as they chased after her to take pictures. She also made the toll takers look bad:

> "Apparently, this turkey decided to make Jersey City her home, alongside of one of the top five busiest toll roads in America," said turnpike spokesman Joe Orlando. "She didn't want to leave, she was a regular, and to be honest with you, she probably had better attendance than a lot of the employees."

Despite this, as is typical, staff had no idea what side they were on. They nicknamed the bird "Tammy"; fed her Cracker Jack; and when she was finally captured and removed, were moved to make bad jokes about missing her. "I think I'm going to have empty nest syndrome," one toll collector said.

Animals can impede the progress of more than traffic and still elicit nothing from the victims except coos and chuckles.

At the Smithsonian's observatory in Arizona, muddy footprints started appearing on their telescopes, not just obscuring the view, but eventually damaging the delicate instruments. You'd think astronomers would react indignantly to this interference with scientific progress and waste of our tax dollars, right? But when the culprit, a relative of the raccoon called a ringtail cat, was found, the reaction was just the opposite:

> "We're considering making the ringtail cat the unofficial mascot of the MEarth project," said project leader David Charbonneau. "With those big eyes, they've certainly got the night vision to be natural-born astronomers!"

EXPERIENCE COUNTS FOR NOTHING

Even people who have every reason to know better are often bad animal enablers. Consider the monkey-phobic British woman who couldn't leave well enough alone. Her fear went back to her childhood when her father raised a chimpanzee, an animal she described as "positively evil." With more close-up and personal experience with nonhuman primates than most people, you'd think she'd realize that she should trust her instincts, right?

No. She decided to go on a trip to Thailand, to tour an island full of macaques and "confront her fears." Here's what happened when she arrived on the beach, laid down her towel, and sat down to observe, as reported in the *Daily Mail*:

> The next thing I noticed, this monkey walked up next to me and I thought, "Oh dear." I began to stand up to move away.

ANIMAL LOVERS

The people who climb over fences to pet zoo animals no doubt hold a variety of misconceptions, including not just the idea that cute animals are sweet and friendly, but also one of my favorites, that animals instinctively know that you're not going to hurt them, a delusion that we'll see more of in the next chapter.

They'd likely all call themselves animal lovers too, but with friends like these, animals don't need enemies, because often it's the animal who pays a much higher price. When a woman climbed a fence and put her arm through another fence to pet a sleeping wolf at the Brookfield Zoo in Chicago, the animal bit her arm and wouldn't let go, so a zoo police officer shot it dead. In another case, a nine-year-old girl who climbed onto a rock ledge and reached over a Plexiglas barrier was bitten by a meerkat at the Minnesota Zoo. All five animals in the exhibit were euthanized to be tested for rabies, since they couldn't be sure which one had bitten her, and the girl's parents didn't want her to have to undergo rabies shots if they weren't necessary.

Unfortunately, she'd waited too long to have second thoughts. The monkey pounced, sank his teeth into her arm, and hung on. Then, backup arrived:

> Then another, bigger monkey bit my arm, just next to the other one biting me, and all of a sudden I was surrounded by monkeys.

From her hospital bed, the woman tried to shift the blame onto the tour operator, but he responded, reasonably, "We can't control the monkeys if they decide to bite someone, that's why we always warn the tourists."

The prize for enabling bad behavior, though, probably has to go to the group of victims of shark attacks who lobby for shark conservation. These people, some of whom are missing limbs, have testified before the U.S. Senate and at the United Nations in favor of prohibiting fishermen from slicing off a shark's fin and throwing the animal back in the water. A barbaric act, indeed—if only the sharks had the same consideration as these extremely forgiving victims.

ONLY TRYING TO HELP

Along with their misguided urges to cuddle and feed and enable, many people feel the need to help allegedly helpless innocent creatures in distress. Some of the clueless find this desire so overwhelming that they provide assistance that is worse than unnecessary. Many a perfectly contented outdoor cat has found himself scooped up from his daily route and confined, then seen his photo plastered on "found cat" posters, when all he wanted was to be left alone to get home in time for dinner.

Some do-gooders take it further by trying to help animals that aren't even animals. Animal control officers regularly respond to calls such as these, taken from reports from local jurisdictions by the *Washington Post*:

> BOWIE, Sept. 5. A resident reported seeing an "injured animal" in the yard of a recently burned-out house on her

block. . . . An animal control officer arrived and determined the "injured" creature was actually a stuffed animal.

UPPER MARLBORO, May 4. A motorist reported a "stray black Labrador retriever" lying motionless in the median on Route 301. . . . An animal control officer found the stray black dog to be a woman's black coat with fur trim.

BELTSVILLE, March 15. A resident reported seeing a swan on the side of the road; it appeared to be injured and unable to move. When an animal control officer arrived with a blanket to try to capture it, he realized the "swan" was a white, long-neck, plastic windshield-washer tank that had fallen from a vehicle.

These calls are so common that an animal control officers' Internet message board has a discussion thread titled "Things that turned out to be other things." Officers know better than to express their opinions of citizen animal ignorance in their written reports, but a little of what's behind the official objective phrasing comes out under the cloak of Internet anonymity:

Weeping caller barely able to tell me where the poor egret was hanging off the wire fence next to the freeway. (Plastic trash bag.)

Weeping caller barely able to tell me where the poor injured baby raccoon was in the road. (Stuffed toy bear. With purple ears.)

Dedicated ACO [animal control officer] journeying way up rough country, over stream, through marsh, TOTALLY

WHOSE SIDE ARE YOU ON?

When it comes to making excuses for animals, sometimes naturalists—who really ought to know better—are the worst culprits:

- In Australia, a wombat that attacked a man for twenty minutes, putting him in the hospital, was stopped only via the use of an ax. A spokesman from the Department of Sustainability and the Environment seemed to call for equal sympathy for the distress of the animal: "If it had mange, it would have been suffering a great deal and would be very intolerant to human interference."
- In South Africa, baboons are devastating to the wine industry, stealing tens of thousands of dollars' worth of grapes ready to harvest—they even prefer the more expensive Pinot Noir over cheaper varieties. But when vineyard owners try to drive away the monkeys with rubber snakes and annoying noise, it brings out the tenured monkey-huggers: "The poor baboons are driven to distraction," said a professor from the University of Cape Town's Baboon Research Unit.
- And in England, scientists have proposed that certain invading species have been around for so long that we might as well give up and declare amnesty, granting "ecological citizenship" to the gray squirrels that have driven out native red squirrels and to the rabbits that cause millions of pounds' worth of damage to crops. They even call for the use of politically correct language:

  Objecting to such terms as "American tree-rats" to describe grey squirrels, they said, "Terms like 'alien species' can risk jingoistic or moralistic stances." They suggest that researchers should instead use such neutral terms as "non-native."

soaked, to injured deer. (Same dam' old log that's been there for ten years.)

After a few calls about a poor dog hit by a car and lying in the snow, sent the officer. It was a dirty snow bank! Several calls later, I made him shovel the snow so people would quit calling and tying up the 1 phone line!

LESSON LEARNED

Some of those who make excuses for animals do eventually realize the mistake they're making. But it's remarkable, and disheartening, to see what it takes. A disabled veteran who was savagely attacked by his service monkey—an animal he called his best friend—said it was worse than the war in Vietnam, where he lost an eye. Joseph Hamric gave this harrowing description to a reporter:

> I got hit all over my body. . . . Cut the vein, tore ligaments out of my wrists. I'm pumping blood all over. . . . I'm looking around and saying "well, never thought I'd go out this way. . . ." I'm sitting there thinking I'm going to die.

A murderous assault is enough to put an end to most human friendships, but apparently it's harder to see the truth when the offender is cute and furry. "He's a great monkey. Even though this happened, he's still my baby," the victim said, and here's how he was rewarded for his forgiving nature, less than two weeks later:

> The 7-year-old capuchin monkey went berserk Monday night just after Hamric fed him pork chops, said Hamric's brother, Bill.

The monkey bit off a pinky and landed his owner in the hospital again—*after he fed him pork chops*, no less. Perhaps that's what was the last straw:

> Now, he's ready to either "put Noah down or give him away," Bill Hamric said.

TEN

Ungrateful Beasts

BY NOW IT SHOULD BE CLEAR THAT NOT ONLY ARE WE BLIND to the flaws of animals but we actually work to maintain a state of denial. We're able to ignore evidence, even if it comes in the form of our own blood.

Still, this might not be so much of a problem if our relationship with animals were more of a two-way street. Of course, it would be unreasonable to demand complete equality. If you're going to do something like spend almost £200,000 ($320,000) to build bridges to help dormice cross the highway, like the Welsh government did, you can't expect the little rodents to return the favor in kind.

However, is it really too much to ask that animals at least show a little gratitude? Apparently so. As we'll see in this chapter, they have no appreciation for our deluded devotion, even though we're doing more and more to please them all the time.

CONSERVATION "SUCCESS STORIES"

Long ago, interactions between humans and animals were pretty much a free-for-all legally. But now, we've got laws protecting animals. And you can just guess what that means.

In Italy, bears are now a protected species after decades of declining population. So a bear nicknamed Dino took that as a license to run amuck. Going beyond normal bear bad behavior like raiding chicken coops and stealing honey from farmers' hives, he also killed and disemboweled donkeys and sheep—and shed a radio collar that at least allowed people to track his location, even if they couldn't do anything about his manners.

Meanwhile in the United States, wild turkeys have not only been protected but encouraged, with predictable results. In Massachusetts, turkeys were extinct by the mid-nineteenth century, but instead of counting their blessings, wildlife officials made repeated attempts to reintroduce the species. Now there are an estimated twenty thousand in the state, and no doubt emboldened by this worshipful treatment, the formerly rare and elusive bird sees no reason to stick to the protection of the woods. They've moved into suburbia, resulting in scenes like this one, which occurred in 2007, when a woman parked her car in Brookline and came face to face with a wild turkey:

> *The turkey eyed Jean-Felix. Jean-Felix eyed the turkey. It gobbled. She gasped. Then the turkey proceeded to follow the Dorchester woman over the Green Line train tracks, across the street, through traffic, and all the way down the block, pecking at her backside as she went.*

"It doesn't take much for them to go berserk," a Brookline resident said of the birds. Mobs of turkeys chasing pedestrians and cyclists and blocking traffic were just the start. Further north in Rockport, Massachusetts, in 2009, mail delivery had to be halted to some houses because turkeys were attacking the carriers and chasing their trucks. In one incident, passers-by had to rescue a mailman: "He was trying to wave a bag full of mail at the turkeys as he ran when some folks pulled over to shoo the turkeys away," said a manager at the post office.

Reports of attacking turkeys are now coming in from all over the country:

- Videotape shows turkeys harassing a small child in a suburb in Pennsylvania.
- In Illinois a turkey crashed into a living room through a plate-glass window.
- In Michigan turkeys trapped a man in his truck at a repair shop.
- Washington State reports another car-attacking turkey.
- A flock in Virginia has taken to attacking commuters. One trapped a sheriff's deputy in his car, requiring him to call an animal control officer, who said that when she arrived, the turkey "just looked right at me and wasn't concerned a bit."

These birds have generally escaped unpunished except when their careless overconfidence gets them hit by vehicles. An exception was in Georgia, where wildlife officials finally passed the death penalty on a gang that was terrorizing a neighborhood, blocking traffic, kicking and biting, and chasing people into their homes. "I tried spraying them with a garden hose and chasing them away

continues on next page . . .

with a broom, but they kept coming back," a resident said. "I've been feeling like a prisoner in my own home. I had to look out the window to make sure they weren't out there."

It's the sociopathic irrationality of their criminal acts that is the most terrifying to some. As one victim said, "If you're attacked by a person, there's usually some reasoning, but a turkey has no reason."

RESCUE ME

Some people claim that animals instinctively know when someone is trying to help them. So why can't they be a little more cooperative? Animals running loose on the roadways, like the many cases we saw in Chapter 2, actively resist efforts to remove them from the dangers of playing in traffic, sometimes taking it to extremes. One dog in Santa Cruz was captured only after eighty attempts, and a swan sitting in traffic on a bridge in London ignored attempts to guide it to safety for over an hour.

And when we've finally saved their furry butts from some peril, we can't expect thanks—on the contrary. Consider, for example, the effort to rescue a dog from a raging rain-swollen river in Los Angeles:

> At least 50 firefighters responded to reports that the dog was in the river. For an hour, firefighters stood at the top of the steep, concrete banks, throwing life vests and float rings, hoping the dog would grab on. . . . One firefighter got into the river and tried to catch him, but the dog took off.

A firefighter dangling from a helicopter tried to reach the dog, but when it hovered near, the dog would scramble away to the banks of the river. Still, the rescuers persisted, and when a brave man finally snatched the dog out of the water, what did he get in return?

> Joe St. Georges, a 25-year Los Angeles Fire Department veteran, said he received a "real bite in the thumb" but was otherwise feeling fine.

Another dog expressed his opinion of a similarly involved rescue in an even more insolent manner. Stranded on an ice floe on Lake Erie, this pooch was also whisked to safety by a brave officer dangling from a helicopter—and then, as soon as they got to shore, ran back out on the lake ice and had to be rescued again.

And in China, see what a pig thinks of not just a miraculous rescue, but the adoration of an entire country:

> A pig that survived 36 days buried in the rubble of May's massive Sichuan earthquake has been voted China's favorite animal, but the attention has made him fat, lazy and bad-tempered, state media said. . . .
>
> People come from all over to see the pig at its new home in a museum, the newspaper said, but it was becoming increasingly spoiled and ungrateful.
>
> And the pig is getting fed up with visitors, after initially being quite friendly.

"Now it just blocks the door to its bedroom when there are too many visitors outside. It's been increasingly difficult for us to convince it to open the door."

THEY CAN'T ALL BE FAIRY-TALE ENDINGS

It's a sad fact of life that not all noble rescue missions have happy outcomes, and it's possible that certain unfortunate incidents make animals skeptical of our motives. For example, it wasn't safe for the squirrel on the plane in Chapter 2 to be running around in the wiring above the cockpit, but he can't have been any more pleased with his final destination: Honolulu authorities, fearing it may have been carrying rabies, had the rodent killed.

Other somewhat less dire outcomes may leave a creature feeling ambivalent at best. A male capybara named Boris escaped from a farm park in Scotland and led the free, single life for months till the temperatures started to fall below comfort for a native of South America. The quest for warmth was his downfall when he wandered into the garage of retired businessman David Hammond:

> David's wife Margaret had the washing machine and tumble dryer on.
>
> "He was in there heating his backside," said David, who quickly closed the garage door.

But he had mixed feelings about assisting with the capybara's homecoming: The giant rodent returned to his mate to discover that she'd given birth to a litter of three while he was gone. "I don't know if I've done Boris a favour or not," Hammond mused. We can only hope that Boris came to appreciate being made to do the right thing.

Still, animals ought to appreciate that sometimes when things don't quite work out, it's the humans who get the worst of it. In fact they sometimes make the ultimate sacrifice, even for the lowliest of creatures, as happened to the six people in Egypt who tried to save a chicken that fell into a well:

> An 18-year-old farmer was the first to descend into the 60-foot well. He drowned, apparently after an undercurrent in the water pulled him down, police said. His sister and two brothers, none of whom could swim well, went in one by one to help him, but also drowned. Two elderly farmers then came to help. But they apparently were pulled by the same undercurrent. The bodies of the six were later pulled out of the well in the village of Nazlat Imara, 240 miles south of Cairo. The chicken was also pulled out. It survived.

NOTHING MORE DANGEROUS THAN AN EDUCATED ANIMAL

Animals have just as little gratitude for longer-term efforts to better their lives. A man in China who found a seriously injured monkey took him home, amputated his ruined limbs, gave him medicine, and nursed him back to health. He then made the one-armed, one-legged primate a part of his family. The monkey supposedly performs chores around the house, if you can call it that: Seeing his master crack some eggs to cook, the monkey went into the chicken coop and smashed all the eggs in it. Then, the man said, the monkey saw him slaughter a chicken:

> From then on, whenever it's not occupied, it jumps into the chicken pen, and kills the chickens, no matter how big or small, and tries to pluck them.
>
> His record is nine chickens in one day. The lesson I have learned is to never slaughter a chicken in front of a monkey.

That monkey learned merely by observation, but another man in China put in what must have been enormous efforts to teach some monkeys the martial art of Tae kwon do and give them a career. Bad idea:

> Lo Wung, 42, taught the monkeys so they could entertain crowds outside a shopping centre in Nshi, in eastern China's Hubei province.
>
> But the money-spinning primates turned the tables on their trainer when he slipped during a show, with one quick-thinking monkey flooring him with a kick to the head.
>
> Hu Luang, 32, a bystander who photographed the incident, said: "I saw one punch him in the eye—he grabbed another by the ear and it responded by grabbing his nose. They were leaping and jumping all over the place. It was better than a Bruce Lee film."

One last story of this sort is perhaps a different sort of cautionary tale about being careful what you teach an animal: In Kyrgyzstan, a bear on ice skates attacked two people during a rehearsal at a circus, killing one of them. Not to take the animals' side, but if I thought someone was going to make me wear one of those ice skating tutus in public, I might bite their leg off too.

TRAVEL THE WORLD, MEET INTERESTING ANIMALS, AND EAT THEM

As demonstrated in Chapter 7 by the bunnies that emigrated to Australia, if you want to cause the most far-reaching bad behavior, take animals to see the world. Consider the brown tree snake. Humans did this snake a big favor, sometime in the 1940s, by giving them a free ride to Guam, probably in military cargo. There were no predators of snakes on Guam, and no native snakes except a tiny, blind, insect-eating creature, so harmless that everyone thought it was some kind of worm.

So when the brown tree snake arrived, it took advantage of a free all-you-can-eat buffet of defenseless wildlife. The innocent creatures had no idea what a snake was and no suspicion that it saw them as food. The snakes proceeded to wipe out most of the delicious bird species, tasty fruit bats, and scrumptious little lizards.

You'd think they'd at least be nice to us humans for providing this great new home, but not so much. Among other problems, they're constantly climbing power poles and causing short circuits, as the U.S. Geological Survey (USGS) explains:

> This results in frequent losses of power to parts of Guam and even island-wide blackouts. Such power failures, brownouts,

> and electrical surges, occurring on average approximately one every three days, damage electrical appliances and interrupt all activities dependent on electrical power, including commerce, banking, air transportation, and medical services. Power outages caused by snakes have been a serious problem on Guam since 1978, and the incidence of snake-caused outages continues to cause significant problems. Records show that more than 1,600 snake-caused outages occurred from 1978–1997.

The USGS quotes an estimate of $4 million per year for research and control of the brown tree snake, including searching outgoing aircraft for stowaways that might invade other islands, and that's not counting damages and losses from blackouts, extinction of wildlife, and environmental problems. And, adding insult to injury, now the snakes are expecting humans to provide food. Having nearly driven all those tasty animals to extinction, larger snakes are now scavenging garbage and even stealing hamburgers off barbecues.

Elsewhere, naturalists are trying to figure out how to solve the problems caused by traveling mice. Ordinary house mice that ships brought to Gough Island in the South Pacific have spent the last century and a half evolving in their new home: They're now two or three times larger and have developed into bloodthirsty carnivores that prey on the chicks of endangered seabirds. The bird babies weigh over twenty pounds compared to the "giant" one-ounce mice. "It is like a tabby cat attacking a hippopotamus," says one naturalist. But the albatrosses and other ground-nesting birds on the island evolved with no native predators, so they have no instinct to defend themselves or their offspring from the rodents that are now eating something like 60 percent of the helpless chicks alive in their nests. Guilt-ridden British conservationists are now struggling to save the colony with plans like dropping poison from helicopters at a projected cost of £2.6 million ($4 million).

And while individual animals can't cause the kind of damage that a whole species can, don't expect them to be any more thankful for the opportunity to travel. The first cat to attempt to cross the Atlantic by air, a tabby called Kiddo, didn't much appreciate the chance to make history. As the airship was being towed out to sea, the wireless operator said, "This cat is raising hell—I believe it's going mad." The crew eventually agreed to try to send the cat back, with the result that the historic first radio communication from an aircraft in flight was:

> Roy, come and get this goddamn cat.

The crew was unsuccessful at lowering Kiddo in a sack into a motorboat, but the attempt caused the cat to reconsider his options—as one crew member said "he suddenly discovered that he could have been in a worse place than an airship"—and he settled down for the rest of the trip, 71.5 hours until the airship came down and all were rescued off the coast of Maryland.

ABOVE THEIR STATION

Despite all this evidence that doing something nice for animals just brings out the worst in them, we're treating them better and better all the time. In fact, they're enjoying more and more of the benefits of civilization that used to belong exclusively to humans. And I'm not just talking about little dogs getting to dress up in fashionable outfits. For instance, in these days of high unemployment, we don't even have enough jobs to go around for our own species, and yet, we're giving jobs to animals.

This trend has gone furthest in Japan. In one case, a cat was given the post of stationmaster at an otherwise unstaffed train station. "Tama is the only stationmaster, as we have to reduce personnel costs," admitted an official. The cat, who did nothing more than wear a uniform cap and watch customers come and go, nevertheless rose rapidly in her career. After only a fairly brief tenure, Tama was given her own office and a promotion to "super-stationmaster," at which point it was reported, appallingly, that she was the only female of any species to hold a managerial position in the railroad company.

Elsewhere in Japan, monkeys wait tables in a traditional sake bar north of Tokyo. They wear traditional Japanese garb, hand out hot towels for hand-cleaning and take drink orders. Customers claim that the monkeys understand their orders. "We called out for more beer just then and it brought us some beer," said one patron. "It's amazing how it seems to understand human words."

These primates are not just taking people's jobs, they're making human employees look bad. "The monkeys are actually better waiters than some really bad human ones," said another customer.

Perhaps most shocking is the fact that this trend is so entrenched in Japan that there are already laws regulating animal employment—and the regulations make their lives a lot easier than those of the human staff, restricting shifts to two hours a day. (Although it seems that these rules do not extend to outlawing child labor: Another train station, impressed by the tourists attracted by Tama the cat, appointed two baby monkeys to their stationmaster position.)

POSITIONS OF RESPONSIBILITY

The Japanese are not the only ones to be guilty of employing nonhuman primates, and in other countries, they've been given jobs with even more authority. There's a monkey on the police force in Thailand that works at a police checkpoint. He allegedly helps defuse tensions with Muslim separatists, although reports provide no explanation of what experience qualifies him for this delicate task. In New Delhi, some companies and government agencies employ langur monkeys as a security force. They're supposed to drive away the troublesome gangs of macaques that run rampant in the city, taking advantage of the fact that devotees of the monkey god Hanuman consider them sacred:

> The gray-brown, pink-faced macaques . . . reportedly have invaded homes and offices, swiping cell phones and sodas, biting children and slapping women. There are stories of them breaking into police stations, donning guns and holsters, and raiding hospitals, where they attack doctors and snatch IVs from patients' arms to slurp the sugary liquid. . . . Everyone knows the story of the monkey who

allegedly mooched booze from a central-Delhi liquor store—Aristocrat vodka and McDowell's whiskey were his favorites.

Using monkeys to fight monkeys may sound only fair, but unfortunately, hiring standards for the anti-macaque force are apparently not very high. Among the monkeys assigned to increase security around the 2010 Commonwealth Games, at least one misused his position to try to steal a BBC cameraman's phone, sending the man to the hospital for treatment of a cut and rabies and tetanus injections.

Sometimes prominent and influential positions even go to animals who have already demonstrated the worst sort of bad behavior. A kakapo in New Zealand gained a certain worldwide notoriety when he sexually assaulted the head of a zoologist who was filming a BBC wildlife documentary. Like some humans, the endangered bird has parlayed this appalling reality-TV moment into a career: It's actually been given a government job as a spokes-parrot for conservation.

THEY REALLY WANT TO DIRECT

As if there weren't enough competition for employment in creative fields these days, animals are also pursuing careers in art. Captive critters that produce paintings have become too numerous to list, including sea lions at an aquarium in Oregon, a rhino at the zoo in Denver, and many elephants. And even those animals—who paint properly with brushes—are probably miffed at what other creatures pass off as art: There are zoos selling works made by penguins and meerkats that do nothing more than walk across the paper with paint on their feet.

Animals have also moved into more modern forms of artistic self-expression. Chimps at a zoo in Scotland were given cameras to film a documentary for the BBC, and an orangutan at the Vienna zoo takes photographs that have gained her tens of thousands of fans on Facebook. Skeptics have attempted to downplay the efforts of these creatures, but with little success. The director of the Vienna zoo, bemoaning the media coverage of their orang photos, told *National Geographic*, "There is no creative touch, no artistic approach!" Explaining that the orangutan manipulates the camera because a raisin is dispensed when she pushes a button, he said, this is just how orangutans are: "We knew that they use and manipulate every object they touch. If you give them a machine-gun, they will soon find out how to shoot it."

Such myth busting has apparently made very little headway, though. A French filmmaker produced a movie filmed and directed by a capuchin monkey, and commented favorably about what he learned working with primates:

> They taught me something very important and very scary for my career: even a well-trained monkey could do my job. Maybe film-makers will soon compete for jobs against monkeys.

SLIPPERY SLOPE

One thing has to be said for those animals, though: At least they're working. After all, it's about time zoo animals really earned their keep, instead of spending most of the day hiding when you're trying to see them.

But once you give animals jobs, where do you draw the line? Is anything in our culture so sacred that they can't participate? Some no longer see a problem inviting them to weddings—and we're not just talking about the increasing number of pet dogs walking down the aisle with bridal parties. In Australia, a pet kangaroo was a bridesmaid, and in Montana, a couple had a bear as their best man. And the next thing you know, animals are getting their own nuptial ceremonies:

- Two pet monkeys in India had a wedding attended by over three thousand guests. The bride was dressed in a red sari, and the primates were showered with gifts including a gold necklace, then released from captivity to lead their married life. One of their former owners said, "I feel as if my own daughter is getting married. I cannot bear the thought that she would not be with us anymore."
- A pair of gay penguins enjoyed a privilege denied to humans in many places when they tied the knot at a zoo in China, wearing traditional red wedding garb and processing to the music of the "Wedding March."

And a dog caused a kerfuffle by receiving communion at a church in Toronto. One parishioner was so shocked that he filed a complaint and left the church. Yet, ominously, others didn't see it as a problem, including this church official who made excuses for the reverend:

I think it was this natural reaction: here's this dog, and he's just looking up, and she's giving the wafers to people and she just gave one to him. Anybody might have done that. It's not like she's trying to create a revolution.

ELEVEN

Heroic Humans

IT'S TIME TO FACE THE TRUTH. YOU KNOW NOW WHAT THESE creatures get away with because they're cute and beautiful and magnificent examples of the wonders of nature. You've seen the problems that result from believing that animals are noble, honest, and good despite all the evidence to the contrary. And you may ask yourself, what can you do to help, aside from buying a copy of this book for everyone you know?

Perhaps the first step is to recognize some role models, like these brave individuals who refused to take bad animal behavior lying down:

- A gardener mowing grass in a city park in India was bitten by a snake. When he couldn't shake it off, he removed it with a pair of scissors, and then he not only bit it back, but chewed up two-thirds of it. He only stopped when be began vomiting and fainted: "I was angry when the snake bit me on my finger. I bit it back because that was my way of taking revenge," Ramesh told doctors after regaining consciousness. His condition is stable. The snake is dead.

- Another "man bites snake" story comes out of Africa: A Kenyan man accidentally stepped on a thirteen-foot-long python and found himself encoiled in its grip and dragged up into a tree. Over the course of three hours, he smothered the snake's head with his shirt to prevent it from swallowing him, bit it on the tail, and successfully struggled to free one arm and get his phone out of his pocket to call for help. Once he was rescued, the snake unfortunately escaped, but a police officer said they were taking the search with the utmost seriousness: "We want to arrest the snake because any one of us could fall a victim."

- In British Columbia, a man saved himself from a bear attack by making do with what was at hand. Too far from his truck to outrun the charging black bear, he looked down, picked up a rock, and drew on his experience as a baseball pitcher, hitting the animal right between the eyes. "It was like I shot it. Knocked it right out," he said.

- A woman in Montana also thwarted a bear attack with an improvised weapon. She attempted to defend her elderly dog from the bear by kicking it, and then, as the animal thrust its head and shoulder through the partly closed door and tried to muscle its way into her house: "The woman held onto the door with her right hand. With her left, she reached behind and grabbed a 14-inch zucchini that she had picked from her garden earlier and was sitting on the kitchen counter. . . . She threw the vegetable. It bopped the bruin on the top of its head and the animal fled." Authorities looking for the bear planned to use DNA retrieved from the zucchini to confirm its identity.

- Perhaps the bravest and most resourceful of all of these—and the one with the best attitude—is the woman who was attacked by a shark while snorkeling and escaped by punching it: "I thought 'this shark's not going to get the better of me' and I started punching it on the nose, punching, punching, punching. . . . And then it got me under the water, but not much because I started kicking at its neck." She lost quite a lot of blood and is going to have to undergo a number of surgeries, but this woman sees a bright side. Sometimes there are unanticipated rewards for those who stand up to bad animals: "I have to have a new remodelled bottom, so that's a positive," she said.

UNITED WE STAND

Individual action against bad animals can't be our only defense, though. We need to pull together as a society. To start, private organizations don't have to depend on the police to regulate animal behavior on their own property. In England, a squirrel was banned from sneaking onto a roller coaster:

> Workers noticed it riding the revamped Sonic Spinball roller coaster as it was tested in the mornings and joining visitors who were offered an early go on it before the official opening.

The gray-haired animal was also caught stealing food from the workers.

A spokesman for the Staffordshire theme park said: "It was getting in the way of builders who were painting. They couldn't carry on because they would end up with paw prints in the paint."

Alarms were installed that emit a warning noise inaudible to human ears but designed to ensure the squirrel, nicknamed Sonic, avoids the ride in future.

We can take heart that despite the fact that amusement parks attract a demographic that is particularly fond of cute furry creatures, there have been no reports of children boycotting this park in response.

And a group of citizens of Rock Hill, South Carolina, set a high standard for grassroots community action in an incident that began when an ordinary man encountered an out-of-the-ordinary creature: "Sure enough there was an ostrich runnin' right by my nose so I couldn't just sit by and do nothin'," house painter Jerry Gibson told a reporter.

The large flightless bird was actually an emu, but wisely, no one stopped to consult a reference book. Rather, neighbors leaped to assist, or at least encourage, as Gibson tried but failed to lasso the bird with an electrical cord.

If you've never been up close with an emu, don't downplay the nerve these folks demonstrated. The enormous birds have got feet like a dinosaur and an attitude to match.

And, showing that it's never too late to join the fight against bad animal behavior, the hero of the day was a seventy-year-old:

> The chase added dozens of Pied Piper–like chasers including the cops, and culminated with the emu's capture by a veteran emu wrangler who just so happens to be a senior citizen with Popeye forearms named Bobby Mangrum. Mangrum was armed with nothing but a fishing net and a bellyful of courage.

He tackled the bird, called for help, and held it down until an officer came and tied it up with a dog leash. See what we can do when we work together?

THE LAW WON

It's only fair to recognize that there are occasional cases where public safety officials set a good example:

- In Germany, a goat was jailed after blocking traffic by standing in the middle of a road and leading police on a chase across town. News items described the animal's bread and water diet and photos showed suitably Spartan accommodations. And the offending creature was insulted, as well: It was reported that the only remarks on the arrest papers were "smells very bad."
- In Columbia, a parrot was taken into custody by police for being part of a conspiracy to sell drugs. The bird, called Lorenzo, was caught trying to warn a local drug cartel of an undercover raid:

  > Lorenzo caused quite the stir as he was presented to journalists. The well-trained creature even showed

off his lookout skills as he yelled out: "Run, run you are going to get caught."

Four men and two other birds were also arrested in the raid, but perhaps the most shocking—and impressive—fact is that Lorenzo is merely the tip of a psittacine criminal iceberg: Authorities claim to have seized over seventeen hundred similarly trained parrots.

- In one part of India they've taken a hard line approach to the widespread problem of marauding monkeys. If you're a primate who wants to make trouble, don't do it in northern Punjab, where you'll pay the price: Hardened criminals get life sentences in "well-guarded and heavily-barred cells" at a local zoo:

  > "All 11 monkeys are hard cases who have been apprehended by game wardens for thieving, terrorising and biting people," their jailer, Ram Tirath, said. "It's unlikely that any of them will ever be paroled."

- The baboons of South Africa whose crimes we noted in Chapter 2 face an even harsher sentence. Authorities have recently adopted a three-strikes policy, and repeat offenders now face the death penalty, although it's a point of controversy:

  > Fourteen-year-old William, a large male known officially as GOB03, who had terrorised the coastal suburb of Scarborough for as long as anyone can remember, was the first to fall foul of this controversial rule.

His death last month was greeted with outrage and jubilation in equal measure and dominated the letters pages of the local newspapers for weeks.

- The punishment fit the crime in the case of Winston, a dog who attacked several vehicles and tore the bumper off a Chattanooga police car. A judge agreed to drop a "potentially dangerous dog" citation after six months if Winston was successful in court-ordered obedience training. The dog's lawyer said, "The obedience training is going to be more like anger management."

GOVERNMENT IN ACTION

In some cases, even a few brave politicians are taking a stand against bad animals.

One admirable example of political leadership comes from the president of Zambia, who was peed on by a monkey while speaking at a press conference. Adopting a zero-tolerance approach, he evicted all two hundred monkeys who'd been enjoying the good life on the grounds of his official residence.

Even more encouraging is the case of an entire electorate that stood up to the tide of unearned privileges being granted to animals. In 2010, Swiss voters rejected a proposal that animals be provided free lawyers by the government. This is all the more noteworthy when seen in context: The Swiss legal system already mandates the mollycoddling of pets, including laws that forbid flushing goldfish down the toilet and require that guinea pigs and canaries be kept with roommates.

Thankfully, there does seem to be a line that even the Swiss won't cross.

SCIENCE SHOWS THE WAY

While we've seen that in some cases, professionals like wildlife biologists can be egregious offenders when it comes to making excuses for bad animal behavior, it's important not to tar all scientists with the same brush: They've come up with some of the most fiendishly clever offensive strategies against our fellow creatures. So, to inspire you to go out and stand up for your species, I'll end this book with a few examples of the most satisfying type: What could be better than using bad animals against bad animals?

GETTING GOATS' GOAT

On an island off South Australia, a population of feral goats, descended from animals brought by settlers for their meat and milk, is following the usual routine of invasive species the world over, and particularly in Australia. They're eating their way through the native plants, including one that's the main food source for the endangered black cockatoo, and basically ruining it for everyone else.

By itself, that's hardly unusual enough to be worthy of mention. The really interesting bad animals in this story are the ones being used by naturalists in the eradication effort: They call them "Judas goats." Goats shipped in from the mainland are sterilized and fitted with radio tracking

SHIFTING THE BLAME

On occasion a human will try to blame an animal for a crime that, for a change, they probably actually didn't commit. This could be a fair way of evening the score for everything else these beasts get away with. But if we're going to use this strategy, we need to do a better job of it. In one case, a woman who was charged with stealing money from her ex-husband's bank account said she had no choice—her dog had gotten into her purse and eaten all her checks. Her story did not persuade prosecutors. In another, a man in Florida was arrested for having child porn on his computer, and tried to blame it on his cat:

> *He told detectives that he would leave the room sometimes while downloading music and when he returned, his cat had jumped up on the keyboard and strange images were downloaded.*

Anyone who has a cat and a computer can imagine that this could happen once; however, the man was in possession of one thousand of these images. He was taken to jail and the cat was not charged.

collars. Then they're released, and—as the leader of the project explains—their natural inclination to join a group takes over:

> "Generally the goat will have about a week's sulk and once it gets over its week of sitting down and just adjusting itself to the new environment it's been released into, it will

NOT THAT LONG AGO

It may be encouraging to know that delusions about our fellow creatures are not a necessary part of the human condition. It wasn't always like this, and I'm not just talking about primitive times when we had to share our homes with livestock and had animal behavior up in our faces on a daily basis. In fact, a more realistic view—one that was even government supported—can be found as recently as the 1940s in a book called *Who's Who in the Zoo: A Natural History of Mammals*, produced by the WPA Federal Writers' Project.

Unconstrained by modern notions of political correctness, this book happily tells readers which animals are used for their fur, desired by big-game hunters for trophies, and good to eat. And the authors have no qualms about revealing unpleasant truths about animal behavior and even insulting them when it's clearly deserved. Here are just a few instructive excerpts:

- Other species of South American monkeys are less surly in captivity than the Howler.
- When a Marmoset is mischievous a slap will not cause it to behave, but it quickly obeys when its ears are pinched or bitten.
- Domesticated (Indian) elephants are used to capture the wild ones. Two tame elephants will squeeze a wild one between them, holding until their masters have bound its legs with chains.
- The Babirusa is one of the ugliest of the wild swine.
- The Guanaco is so stupid that the native Patagonian Indians are able to surround the herds and club many of their members to death.

- The mother [tiger] rarely deserts the young in infancy, unless hard-pressed. But she has been known to eat her kittens when food was scarce.
- The Camel is known to have served man for the last 5,000 years, but despite long domestication it has a very ugly disposition and is not attached to its master.

One also has to admire the writers' skill at getting in a dig at large groups of animals while weakly complimenting one of them, as we see in this remark about the capybara:

- Largest of all living rodents, the Capybaras are the least obnoxious.

That seems pretty fair and balanced, right? Why can't we see more of this in nature writing today?

go off . . . and find friends," he said. "The feral goats are quite happy to accept them into their mobs and they fit right in."

The radio collars allow officials to find the group of feral goats and then, let's just say, they don't gather them up and take them to a goat shelter and adopt them out. And what the Judas goats don't know is that they're going to pay the price for their collusion: Once the feral goat population has been eliminated, the Judas goats will follow.

HOP TO IT

One of the most famous invasive bad animals in Australia and perhaps the world is the cane toad. Farmers brought cane toads to Australia in the 1930s, and, in exchange for a whole new continent to live on, they asked only that the toads eat a certain beetle that infests sugarcane fields. Seems like a fair deal, right?

Instead, the toads pretty much ignored the beetles and have spread across the country, eating native species and out-competing them for food. The toads also add their own special touch to this familiar routine: They produce venom in warty lumps in their skin, which sometimes kills animals that pick up toads in their mouths.

Years of attempts to control the invading amphibians have failed, but now scientists may have found a simple solution. It's an extremely nasty little insect, the carnivorous meat ant, that finds cat food delicious—and also won't say no to a nice meal of baby toads:

> "It's not exactly rocket science. We went out and put out a little bit of cat food right beside the area where the baby toads were coming out of the ponds," University of Sydney professor Rick Shine told public broadcaster [Australian Broadcasting Corporation].
>
> "The ants rapidly discovered the cat food and thought it tasted great."
>
> "The worker ants then leave trails back to the nest encouraging other ants to come out there and forage in that area, and within a very short period of time we got lots of ants in the same area as the toads are."

Along with the cat food appetizer, the ants gobble up about 70 percent of the toads immediately, and most that escape the attack die later as well. Targeting the toads as they hatch is a particularly efficient approach, since the eggs are laid in huge masses and tens of thousands of young emerge at the same time.

It's a particularly beautiful story of payback. Immigrant species usually take advantage of the fact that the natives have no defenses against them. But here, the natives are perfectly set to work together against the toads: The indigenous frogs have evolved the ability to dodge the ants, so they're unaffected by the cat-food-induced attacks.

THROWING CUTE RIGHT BACK AT THEM

Finally, the research that has perhaps the most wide-ranging implications comes out of Japan, where Japanese macaques hide behind a reputation as the cute, fuzzy snow monkeys that wash their food, play with snowballs, and take blissful, relaxing baths in hot springs. In reality they behave just as badly as their relatives in other parts of the world. In the summer of 2010, more than sixty people in towns near Mount Fuji were the victims of monkey attacks. The most vulnerable citizens were not even safe in their own homes: An eighty-year-old woman was attacked from behind on her porch, and a first-grade girl was scratched by a monkey that entered her house.

However, scientists recently discovered that getting back at the monkeys is simple. All you have to do is show them a cute, inoffensive flying squirrel:

> When Japanese giant flying squirrels glided over to a tree in the monkeys' vicinity, adults and adolescent macaques started

hollering at it threateningly, the researchers report. Young macaques screamed and mothers scooped up their infants, while adults and high-ranking males in particular went and physically harassed the offending squirrel.

The squirrels certainly can't hurt the much larger monkeys, and they don't compete very much for food resources. The scientists speculate on the underlying cause of the reaction, but whatever it is, the monkeys just can't stand the creatures. The eminent *Christian Science Monitor* summed up this story perfectly:

> **MONKEYS HATE FLYING SQUIRRELS, REPORT MONKEY-ANNOYANCE EXPERTS**
>
> Japanese macaques will completely flip out when presented with flying squirrels, a new study in monkey-antagonism has found. The research could pave the way for advanced methods of enraging monkeys.

The importance of this research goes far beyond these pesky primates, however. Merely maddening monkeys is only a first step. If science can infuriate and annoy macaques, there's nothing to stop them from developing ways to irritate dolphins, exasperate bears, and piss off pandas. And you don't need an advanced degree to play a role: Anyone can put this book down right now and discombobulate a dog, aggravate a cat, outrage a goldfish, or perturb a squirrel.

So visualize a world where we face the facts about animals: Nature is nasty, and it's more interesting that way. If you've read this far, you're a member of the precious minority who understands that animals aren't as cute as they want you to think. Stand up for the truth, and together we can stop them from pulling the wool, fur, feathers, and scales over our all-too-willing eyes.

ACKNOWLEDGMENTS

Thanks for contributing their expert services to the fight against bad animals on my blog and in this book are due to:

Molly Diesing, German-language correspondent
David Feldman, virtual reference librarian
Alison Taub, flat-coated retriever specialist
Robin Saunders, consulting herpetologist

Thanks also to my agent, Myrsini Stephanides, and editor, Meg Leder. And as always, special thanks to Technical Staff for Life, Ron Boucher, and to Beth Harpaz, who once said, "You should write a book."

REFERENCES

INTRODUCTION: NOT AS CUTE AS THEY WANT YOU TO THINK

Angier, Natalie, "A Highly Evolved Propensity for Deceit," *New York Times*, Dec. 22, 2008, www.nytimes.com/2008/12/23/science/23angi.html.

"Hungry Thai Elephants Raid Villages, Hijack Sugarcane Trucks," AFP/*Sydney Morning Herald*, Dec. 9, 2003, www.smh.com.au/articles/2003/12/08/1070732147207.html.

Kemp, Joe, Adam Lisberg, and Helen Kennedy, "How's This for a Warning? Groundhog Staten Island Chuck Bites Mayor Bloomberg on Big Day," *New York Daily News*, Feb. 2, 2009, www.nydailynews.com/ny_local/2009/02/02/2009-02-02_hows_this_for_a_warning_groundhog_staten-1.html.

"'Ladykiller' Cockerel Left Lonely After Mate Dies from Exhaustion," *Telegraph* (UK), July 9, 2009, www.telegraph.co.uk/news/newstopics/howaboutthat/5782620/Ladykiller-cockerel-left-lonely-after-mate-dies-from-exhaustion.html.

"MP Blunkett Injured in Cow Attack," BBC News, June 8, 2009, http://news.bbc.co.uk/2/hi/uk_news/england/8089498.stm.

Savage, James, "Boozing Bunny Blessed with Four Lucky Feet," *Worcester News*, Jan. 23, 2010, www.worcesternews.co.uk/news/4867260.Boozing_bunny_blessed_with_four_lucky_feet.

1. MUGGERS, BURGLARS, AND THIEVES

"Baboons Steal Underwear from Rooftop Luggage," *Telegraph* (UK), July 20, 2009, www.telegraph.co.uk/news/uknews/5870659/Baboons-steal-underwear-from-rooftop-luggage.html.

Beard, Matthew, "The Curious Incident of the Hungry Dog in the Night-Time," *Independent* (UK), Oct. 5, 2004, www.independent.co.uk/news/uk/this-britain/the-curious-incident-of-the-hungry-dog-in-the-nighttime-535642.html.

Bolton, Ash, "Southampton Cat Keeps Stealing Underwear from across Portswood," *Southern Daily Echo* (UK), July 8, 2010, www.dailyecho.co.uk/news/8261256.Cat_burglar_stealing_underwear.

Coyle, Meg, "Feline with a Fetish Goes for Gloves," King5, July 14, 2010, www.king5.com/news/Glove-Stealing-Cat-98464489.html.

"Criminal Birds: Brazen Magpie Looting German Village," Spiegel Online International, Jan. 3, 2008, www.spiegel.de/international/zeitgeist/0,1518,526445,00.html.

"Dog Eats $400, but Woman Recovers Some Cash Later," AP/*Los Angeles Times*, March 18, 2009, www.latimes.com/entertainment/la-on-dogmoney19-2009mar19,0,351964.story.

"Dog Swallows $20,000 Diamond in Maryland Jewelry Store," WJLA, March 11, 2010, www.wjla.com/news/stories/0310/715111.html?ref=tw.

"Drug-Smuggling Pigeon Arrested," AFP/*Brisbane Times*, Aug. 22, 2008, www.brisbanetimes.com.au/news/world/drugsmuggling-pigeon-arrested/2008/08/22/1219262454395.html.

Flanagan, Jane, "Drunk Baboons Plague Cape Town's Exclusive Suburbs," *Telegraph* (UK), Aug. 29, 2010, www.telegraph.co.uk/news/worldnews/africaandindianocean/southafrica/7969313/Drunk-baboons-plague-Cape-Towns-exclusive-suburbs.html.

"Frankie the Feline Exposed as the Cat Burglar Stealing Toys from His Neighbours' Homes," *Daily Mail*, Dec. 9, 2008, www.dailymail.co.uk/news/article-1093194/Frankie-feline-exposed-cat-burglar-stealing-toys-neighbours-homes.html.

French, Brett, "Bear Break-Ins: Red Lodge Kitchens Raided by Elusive Bruin," *Billings Gazette*, Aug. 5, 2010, http://billingsgazette.com/news/state-and-regional/montana/article_4a3788d4-a0e5-11df-855f-001cc4c03286.html.

Gast, Phil, "Bear Gets Stuck in Car, Goes on Brief Ride," CNN, July 24, 2010, www.cnn.com/2010/US/07/23/colorado.bear.car/index.html.

Hotton, Mark, "Cheeky Kea Steals Tourist's Passport," *Southland Times*, May 29, 2009, www.stuff.co.nz/national/2455068/Cheeky-kea-steals-tourists-passport.

"'Imelda' Strikes Again: Thieving Fox Amasses 120 Shoes," Spiegel Online International, June 10, 2009, www.spiegel.de/international/zeitgeist/0,1518,629778,00.html.

"Killer Chipmunks Invade Kitchens," *Sun* (UK), July 27, 2009, www.thesun.co.uk/sol/homepage/news/2556542/Killer-chipmunks-invade-kitchens.html.

Lazzeri, Antonella, "Mum Attacked by Killer Chipmunk," *Sun* (UK), July 30, 2009, www.thesun.co.uk/sol/homepage/news/2562305/Mum-attacked-by-killer-chipmunk.html.

Lazzeri, Antonella, and Andy Whelan, "Chipmunch," *Sun* (UK), July 28, 2009, www.thesun.co.uk/sol/homepage/news/2558075/Killer-chipmunk-invades-kitchen.html.

Marshall, Leon, "Baboon 'Gangs' Run Wild in Suburban South Africa," *National Geographic News*, Oct. 4, 2006, http://news.nationalgeographic.com/news/2006/10/061004-baboons.html.

Marshall, Tim, "Spy Pigeons Coo in Iran," *Sky News*, Oct. 22, 2008, http://blogs.news.sky.com/foreignmatters/Post:8c2a2858-01ab-4ab8-9291-42bd4afa79b6.

McKenzie, Bryan, "Aggressive Fox Bites 2 People, Steals Sweater," *Daily Progress*, Aug. 27, 2009, www2.dailyprogress.com/news/cdp-news-local/2009/aug/27/aggressive_fox_bites_2_people_steals_sweater-ar-74829.

"Mischievious Monkey Makes Off with Multiple Spectacles," *Hindustan Times*, Sept. 3, 2010, www.hindustantimes.com/Mischievious-monkey-makes-off-with-multiple-spectacles/Article1-595495.aspx.

"Monkey at large after plant store thefts," UPI, July 22, 2009, www.upi.com/Odd_News/2009/07/22/Monkey-at-large-after-plant-store-thefts/UPI-50651248300569.

Moszczynski, Joe, "Black Bear Knocks Down Vernon Man, Steals Sandwich," *Newark Star-Ledger*, July 1, 2009, www.nj.com/news/index.ssf/2009/07/black_bear_attacks_vernon_man.html.

"Oink, Oink, Oops! Ginger the Greedy Pig Swallows Diamond from £1,500 Wedding Ring," *Daily Mail*, Aug. 12, 2009, www.dailymail.co.uk/news/article-1205723/Naughty-pig-swallows-diamond-1-000-ring.html.

Pankratz, Howard, "Car Thief Turns Out to Be a Bear," *Denver Post*, Oct. 15, 2009, www.denverpost.com/technology/ci_13570692.

Phillips, Rhodri, "Breakout Fuels Killer Chipmunk Concerns," *Sun* (UK), July 27 2009, www.thesun.co.uk/sol/homepage/news/2556293/Breakout-fuels-killer-chipmunk-fears.html.

Sapa, "Baboons Listen for Car Alarms," *Times* (South Africa), July 21, 2010, www.timeslive.co.za/scitech/article562946.ece/Baboons-listen-for-car-alarms.

"Seagull Becomes Crisp Shoplifter," BBC News, July 20, 2007, http://news.bbc.co.uk/2/hi/uk_news/scotland/north_east/6907994.stm.

"Squirrel Stealing Flags at Mich. Cemetery," UPI, May 30, 2009, www.upi.com/Odd_News/2009/05/30/Squirrel-stealing-flags-at-Mich-cemetery/UPI-12891243709217.

Stoler, Steve, "Police Believe Monkey Used to Steal from Business," WFAA, July 21, 2009, www.wfaa.com/news/local/64586287.html.

"Wing Wing: Pigeons Take Mobiles into Brazil Jail," AFP/Australian Broadcasting Corporation, March 31, 2009, www.abc.net.au/news/stories/2009/03/31/2531442.htm.

"World Cup Tourists Warned of Baboon-Jackers," AFP/*Age*, May 5, 2010, www.theage.com.au/travel/travel-news/world-cup-tourists-warned-of-baboonjackers-20100505-u8jv.html.

Yapp, Robert, "Colombian Police Capture Pigeon Carrying Drugs to Prison," *Telegraph* (UK), Jan. 18, 2011, www.telegraph.co.uk/news/worldnews/southamerica/brazil/8269366/Colombian-police-capture-pigeon-carrying-drugs-to-prison.html.

2. ASSAULT, RUNNING AMOK, AND ARSON

Allen, Nick, "Rock Band Kings of Leon Forced Off Stage by Pigeon Droppings," *Telegraph* (UK), July 25, 2010, www.telegraph.co.uk/news/worldnews/northamerica/usa/7909398/Rock-band-Kings-of-Leon-forced-off-stage-by-pigeon-droppings.html.

"Army of Frogs Closes Greek Highway," *Telegraph* (UK), May 26, 2010, www.telegraph.co.uk/news/newstopics/howaboutthat/7767860/Army-of-frogs-closes-Greek-highway.html.

"Attack at SJ's Evergreen School," CBS News, May 10, 2007, http://replay.waybackmachine.org/20090609065838/http://cbs5.com/local/Squirrel.Attack.Evergreen.2.455548.html

"Austrian Woman Reports Otter Attack in Wisconsin," AP/*Seattle Times*, Aug. 11, 2009, http://seattletimes.nwsource.com/html/nationworld/2009642132_apusotterattackwisconsin.html.

Barrowclough, Anne, "Man Wrestles Crazed Ninja Kangaroo After It Invades Family Home," *Times*, March 10, 2009, www.timesonline.co.uk/tol/news/world/asia/article5875004.ece.

Bermant, Charlie, "Cat's 'Step Aerobics' Blamed for House Fire in Port Townsend," *Peninsula Daily News*, Nov. 9, 2010, www.peninsuladailynews.com/article/20101109/news/311099985/cats-step-aerobics-blamed-for-house-fire-in-port-townsend.

Bhusnurmath, Mythili, "Too Much Monkey Business," *Economic Times*, Oct. 28, 2007, http://economictimes.indiatimes.com/Columnists/Too_much_monkey_business/articleshow/2496079.cms.

Blake, Heidi, "Rats Attack the Railways," *Telegraph* (UK), Feb. 3, 2010, www.telegraph.co.uk/news/uknews/road-and-rail-transport/7139748/Rats-attack-the-railways.html.

"Boar Draw—Four-Legged Fans Run Amok on German Pitch," AFP/France24, Sept. 3, 2010, www.france24.com/en/20100903-boar-draw-four-legged-fans-run-amok-german-pitch.

"Boar Rampage—One Went to Church, Another to the Market," Spiegel Online International, March 17, 2008, www.spiegel.de/international/zeitgeist/0,1518,541867,00.html.

Bobo, Jeff, "Cows Licking House Cause $100 Damage," Times-News Online, Dec. 6, 2009, http://timesnews.net/article.php?id=9018881.

"Boy, 3, Hospitalized in Playground Squirrel Attack," WESH, Sept. 5, 2007, www.wesh.com/r/14052662/detail.html.

Chandler, Mark, "Catford: Elderly Dancer Reveals Crow Attack Horror," *News Shopper*, June 9, 2010, www.newsshopper.co.uk/news/8208903.CATFORD__Elderly_dancer_reveals_crow_attack_horror.

———, "Eltham: Crow Attacks Leave Blonde Joggers in a Flap," *News Shopper*, May 28, 2010, www.newsshopper.co.uk/news/8190485.ELTHAM__Crow_attacks_leave_blonde_joggers_in_a_flap.

"Chaos and Destruction in German Town—Invasion of the Wild Boars," Spiegel Online International, Nov. 27, 2006, www.spiegel.de/international/0,1518,450995,00.html.

"Chased by an Angry Squirrel," *Metro*, April 29, 2003, www.metro.co.uk/weird/3416-chased-by-an-angry-squirrel.

"Columbus-Bound Flight Delayed by Otters," 10TV, Dec. 23, 2009, www.10tv.com/live/content/local/stories/2009/12/23/story-columbus-houston-flight-otters.html?sid=102.

Comenetz, Jacob, "Chimp Bites Off Berlin Zoo Director's Finger," Reuters, June 10, 2009, www.reuters.com/article/idUSTRE55957X20090610.

Cox, Teri, "Damage Caused by Runaway Zebras," KTXL, Aug. 16, 2010, www.fox40.com/news/headlines/ktxl-news-damagedcausedbyrunawayzebras-0816,0,2425191.story.

Crossland, David, "Climate Change's Clear Winners: Europe's Wild Boar Population Exploding," Spiegel Online International, Nov. 25, 2009, www.spiegel.de/international/zeitgeist/0,1518,663411,00.html.

Deutsch, Kevin, and Leo Standora, "Bronx Man Finds 3-Foot-Long Corn Snake Coiled Comfortably on His Toilet Seat," *New York Daily News*, Sept. 21, 2010, www.nydailynews.com/ny_local/2010/09/21/2010-09-21_3ft_snake_is_on_my_toilet_seat.html.

Dhillon, Amrit, "Stampeding Elephants Dominate Political Agenda in Indian Elections," *Telegraph* (UK), April 4, 2009, www.telegraph.co.uk/news/worldnews/asia/india/5105256/Stampeding-elephants-dominate-political-agenda-in-Indian-elections.html.

"Disturbing the Dead—Wild Boar Wreck One of Europe's Biggest Cemeteries," Spiegel Online International, Oct. 2, 2007, www.spiegel.de/international/zeitgeist/0,1518,509138,00.html.

"Elephant Holds Up Rail Traffic—Train Engine Damaged," *Telegraph* (Calcutta), Aug. 17, 2010, www.telegraphindia.com/1100818/jsp/northeast/story_12822176.jsp.

"Elephants Loom Large as an Issue in Indian Elections," Reuters, March 31, 2009, www.reuters.com/article/idUSDEL95647.

"Elephants Trample Indian Democratic Process," BreakingNews, April 27, 2009, www.breakingnews.ie/world/elephants-trample-indian-democratic-process-408510.html.

"'Elephant!' SUV Sideswipes Circus Escapee," AP/MSNBC, Nov. 5, 2009, www.msnbc.msn.com/id/33675514.

"Escaped Zebra Captured in Downtown Atlanta," AP/MSNBC, Feb. 19, 2010, www.msnbc.msn.com/id/35469033/ns/us_news-weird_news.

Fene, Deanna, "Snake in Toilet Bites Woman," *First Coast News*, July 20, 2005, www.firstcoastnews.com/news/local/story.aspx?storyid=40891.

"Flying Rabbits, Violent Cows and Drowning Hedgehogs," Spiegel Online International, Aug. 13, 2009, www.spiegel.de/international/zeitgeist/0,1518,642112-4,00.html.

Gentleman, Amelia, "Monkeys in the Parks, Monkeys in the Palace," *New York Times*, Nov. 14, 2007, www.nytimes.com/2007/11/14/world/asia/14delhi.html.

"German Safari: Augsburg Police Spend Hours Chasing Escaped Zebras," Spiegel Online International, Feb. 9, 2009, www.spiegel.de/international/zeitgeist/0,1518,606487,00.html.

"Goat Starts Iowa Fire; Cat Raises Alarm," UPI, Jan. 14, 2009, www.upi.com/Odd_News/2009/01/14/Goat-starts-Iowa-fire-cat-raises-alarm/UPI-55651231954315.

Grady, Denise, "Dangerous Cows," *New York Times*, July 31, 2009, http://tierneylab.blogs.nytimes.com/2009/07/31/dangerous-cows.

Greenberg, Emily, and Toby Sells, "Critter's Mischief Creates Mayhem; Long Power Blackouts Rare in Medical Center," *Commercial Appeal*, June 26, 2010, www.commercialappeal.com/news/2010/jun/26/critters-mischief-creates-mayhem.

Grubaugh, Dennis, "Cat Triggers Outage, Lives to Meow About It," *Alton (IL) Telegraph*, Nov. 23, 2009, www.thetelegraph.com/news/power-33442-carrollton-residents.html.

Hawley, Charles, "A Quarter Century after Chernobyl Radioactive Boar on the Rise in Germany," Spiegel Online International, July 30, 2010, www.spiegel.de/international/zeitgeist/0,1518,709345,00.html.

"Hero in Underpants 'Tackles 'Roo,'" BBC News, March 10, 2009, http://news.bbc.co.uk/2/hi/asia-pacific/7932870.stm.

Hough, Andrew, "Wild Boxing Kangaroo Knocks Australian Jogger Unconscious After Assault," *Telegraph* (UK), March 19, 2010, www.telegraph.co.uk/news/worldnews/australiaandthepacific/australia/7481566/Wild-boxing-kangaroo-knocks-Australian-jogger-unconscious-after-assault.html.

Hunter, Thomas, "Roo Rage: How Rocky and Chris Survived," *Sydney Morning Herald*, Nov. 24, 2009, www.smh.com.au/national/roo-rage-how-rocky-and-chris-survived-20091124-jeoe.html.

"Jackdaw Causes Havoc in Quiet Nottinghamshire Village," *Telegraph* (UK), Oct. 9, 2009, www.telegraph.co.uk/news/newstopics/howaboutthat/6284569/Jackdaw-causes-havoc-in-quiet-Nottinghamshire-village.html.

Jennings, Ralph, "Toilet Snake Attack: Urban Legend Comes True?" Reuters, May 11, 2009, www.reuters.com/article/idUSTRE54A5WL20090511.

"Kangaroo Attacks," Amazing Australia, www.amazingaustralia.com.au/animals/kangaroo_attacks.htm.

Kim, Christine, "South Korean Plane Delayed by Bird on Board," Reuters, Oct. 7, 2009, www.reuters.com/article/idUSTRE5961A920091007.

Lewis, Leo, "Tokyo Crows Hack into 'Internest,'" *Times*, June 15, 2006, www.timesonline.co.uk/tol/news/world/asia/article1083581.ece.

"Llama Drama on German Motorway," AFP/France 24, July 14, 2010, www.france24.com/en/20100714-llama-drama-german-motorway.

McGreevy, Ronan, "Traffic Chaos as Circus Llamas Take to the Highway," *Irish Times*, Oct. 2, 2009, www.irishtimes.com/newspaper/ireland/2009/1002/1224255679097.html.

"Monkey Menace: Delhi Deputy Mayor S S Bajwa Dies," *Times of India*, Oct. 21, 2007, http://timesofindia.indiatimes.com/Monkey_menace_Delhi_Deputy_Mayor_S_S_Bajwa_dies/rssarticleshow/2478340.cms.

"'National Pet Fire Safety Day' Prevention Tips to Keep Pets from Starting Home Fires," American Kennel Club, July 12, 2010, www.akc.org/news/index.cfm?article_id=4152.

"Okla. Electric Crews Say Bobcat Caused Outage," AP/ABC News, March 11, 2010, http://abcnews.go.com/GMA/WaterCooler/wireStory?id=10065514.

"Police Become Prey—Wild Boar Hunts Police in German Town," Spiegel Online International, March 4, 2008, www.spiegel.de/international/zeitgeist/0,1518,539289,00.html.

"Possible Alligator Firestarter Back with Owner; Investigation Continues," WTAE, March 5, 2009, www.thepittsburghchannel.com/news/18863914/detail.html.

Powell, David, "Pet Dog Sets Fire to Kitchen in Llanfairfechan," *Daily Post*, Nov. 11, 2009, www.dailypost.co.uk/news/north-wales-news/2009/11/11/pet-dog-sets-fire-to-kitchen-in-llanfairfechan-55578-25139441.

Pushkin, Yuri, "Ice Skating Bear Kills Russian Circus Hand," CNN, Oct. 23, 2009, www.cnn.com/2009/WORLD/europe/10/23/russia.skating.bear.death.

Ricigliano, Ryan, "VFW Fire? Blame the Flaming Rat," *Yakima (WA) Herald Republic*, June 18, 2010, www.yakima-herald.com/stories/2010/06/18/flaming-rat-caused-vfw-fire.

Sanderson, W. T., "Fatalities Caused by Cattle—Four States, 2003–2008," *Morbidity and Mortality Weekly*, July 31, 2009, www.cdc.gov/mmWR/preview/mmwrhtml/mm5829a2.htm.

Schulz, Matt, and Kelly Ryan, "Goat Attack Leaves Three Injured at On Luck Chinese Nursing Home in Donvale," *Herald Sun* (Australia), April 20, 2010, www.heraldsun.com.au/news/victoria/goat-attack-leaves-three-injured-at-on-luck-chinese-nusing-home-in-donvale/story-e6frf7kx-1225855963381.

Sharp, David, "Motorists Warned to Beware of Moose on the Move," AP/*Boston Globe*, May 28, 2010, www.boston.com/news/local/vermont/articles/2010/05/28/new_england_motorists_warned_to_beware_of_moose.

Shimogawa, Duane, "Threatened Bird Brings an End to Friday Night Lights on Kauai," *Hawaii News Now*, July 21, 2010, www.hawaiinewsnow.com/Global/story.asp?S=12843587.

"Six Foot Long Snake Discovered in Toilet in Poland," *Telegraph* (UK), Sept. 20, 2010, www.telegraph.co.uk/news/newstopics/howaboutthat/8014470/Six-foot-long-snake-discovered-in-toilet-in-Poland.html.

"Skittering Squirrel Forces Plane to Land," AP/*USA Today*, Feb. 15, 2007, www.usatoday.com/travel/news/2007-02-15-skittering-squirrel-forces-plane-landing_x.htm.

"Snake Plunges Philippine Island into Darkness," AFP/Yahoo! News, Sept. 10, 2010, www.google.com/hostednews/afp/article/ALeqM5i1GvFJjvcIvzVQq6KVxbmTXUIYTA.

"Squirrel Attacks Anger Winter Park Residents," Click Orlando, Aug. 11, 2006, www.clickorlando.com/news/9665269/detail.html.

Taylor, Joel, "Villagers Live in Fear of Dive-Bombing Buzzard," *Metro*, June 14, 2010, www.metro.co.uk/news/830702-villagers-live-in-fear-of-dive-bombing-buzzard.

"Trunk and Disorderly: Sabu the Escaped Elephant Sparks Chaos on the City Streets," *Daily Mail*, June 8, 2010, www.dailymail.co.uk/news/worldnews/article-1284727/Trunk-disorderly-Sabu-escaped-elephant-sparks-chaos-streets-Zurich.html.

"Wallaby Goes on Run on Motorway," *Telegraph* (UK), Jan. 18, 2010, www.telegraph.co.uk/news/newstopics/howaboutthat/7012129/Wallaby-goes-on-run-on-motorway.html.

Webb, Helena, "Roo Rage Stories Excite Researcher," Australian Broadcasting Corporation, Feb. 3, 2005, www.abc.net.au/wa/stories/s1295464.htm.

"Wild Boar Crash Test Highlights Growing Accident Risk," Spiegel Online International, April 20, 2010, www.spiegel.de/international/zeitgeist/0,1518,690156,00.html.

"Wild Boar Storms Frankfurt Church," Spiegel Online International, Nov. 28, 2008, www.spiegel.de/international/zeitgeist/0,1518,593287,00.html.

"Wild Elephant Causes Traffic Jam," CBS News, Aug. 19, 2010, www.3news.co.nz/Wild-elephant-causes-traffic-jam/tabid/1160/articleID/171538/Default.aspx.

Yancey, Kitty Bean, "Snakes on a Plane Isn't Just a Movie," *USA Today*, Aug. 14, 2006, www.usatoday.com/life/2006-08-14-snakes_x.htm.

"YFD: Rat Caused VFW Fire," KNDO/KNDU, June 18, 2010, www.kndo.com/Global/story.asp?S=12675091.

"Zebras Corraled After Wandering Calif. Streets," AP/Boston.com, Aug. 16, 2010, www.boston.com/news/odd/articles/2010/08/16/zebras_corraled_after_wandering_calif_streets.

3. KINKY CREATURES

Afianos, Jasmin, "Horny Roo Stalks NT Women," NT News, May 14, 2010, www.ntnews.com.au/article/2010/05/14/147401_ntnews.html.

Bagemihl, Bruce, *Biological Exuberance: Animal Homosexuality and Natural Diversity*, St. Martin's Press, 1999 , pp. 69–70, 126, 205, 208, 209, 273, 274, 277–278, 286, 291, 305, 310, 333, 346, 355, 358, 365, 369, 373, 376, 383, 385, 389, 435, 451, 463, 469, 471, 475, 481, 482, 485, 494, 503, 538, 542, 581, 586, 639, 642, 643, 650, 681.

Bakke, Dave, "Goose Won't Be Deterred from Concrete Girlfriend," *State Journal-Register*, March 23, 2010, www.sj-r.com/bakke/x1526478793/Dave-Bakke-Goose-wont-be-deterred-from-concrete-girlfriend.

Barash, David, and Judith Eve Lipton, *The Myth of Monogamy: Fidelity and Infidelity in Animals and People*, Freeman, 2001, pp. 39, 52–54, 89.

Darwin, Charles, *The Descent of Man, and Selection in Relation to Sex*, Clarke, Given and Hooper, 1874, chap. 14.

Deaner, Robert O., Amit V. Khera, and Michael L. Platt, "Monkeys Pay Per View: Adaptive Valuation of Social Images by Rhesus Macaques," *Current Biology* 15, no. 6 (2005): 543–548, www.cell.com/current-biology/abstract/S0960-9822%2805%2900093-X.

Dixson, A. F., *Primate Sexuality: Comparative Studies of the Prosimians, Monkeys, Apes, and Human Beings*. Oxford University Press, 1998, pp. 139, 140, 142.

French, Thomas, *Zoo Story: Life in the Garden of Captives*, Hyperion, 2010, p. 34–35, 143.

Gilbya, Ian C., M. Emery Thompson, Jonathan D. Ruane, and Richard Wrangham, "No Evidence of Short-Term Exchange of Meat for Sex Among Chimpanzees," *Journal of Human Evolution* 59, no. 1 (2010): 44–53, doi:10.1016/j.jhevol.2010.02.006.

Gray, Denis D., "Porn Sparks Panda Baby Boom in China: Research—and Blue Movies—Attributed to Record-High Birth Rate in 2006," MSNBC, Nov. 27, 2006, www.msnbc.msn.com/id/15852885.

Gumert, Michael D., "Payment for Sex in a Macaque Mating Market," *Animal Behaviour* 74, no. 6 (2007): 1655–1667, doi:10.1016/j.anbehav.2007.03.009.

Hrdy, Sarah Blaffer, *Mother Nature: A History of Mothers, Infants, and Natural Selection*, Pantheon, 1999, p. 89.

Hunter, F. M., and L. S. Davis, "Female Adélie Penguins Acquire Nest Material from Extrapair Males After Engaging in Extrapair Copulations," *The Auk* 115, no. 2 (1998): 526–528, www.jstor.org/pss/4089218.

Judson, Olivia, *Dr. Tatiana's Sex Advice to All Creation: The Definitive Guide to the Evolutionary Biology of Sex*, Holt Paperbacks, 2003, pp. 88, 106–108.

Mahr, Krista, "Do Monkeys Pay for Sex?" *Time*, Jan. 7, 2008, www.time.com/time/health/article/0,8599,1700821,00.html.

Marshall, Michael, "The Sex Lives of Animals: A Rough Guide," *New Scientist*, Sept. 2, 2008, www.newscientist.com/article/dn14638-animals-behaving-badly.html?feedId=sex_rss20.

McDonnell, S. M., and A. L. Hinze, "Aversive Conditioning of Periodic Spontaneous Erection Adversely Affects Sexual Behavior and Semen in Stallions," *Animal Reproduction Science* 89 (2005): 77–92, http://research.vet.upenn.edu/Portals/49/SEAM_ISSR_2005.pdf.

Milner, Richard N. C., Michael D. Jennions, and Patricia R. Y. Backwell, "Safe Sex: Male–Female Coalitions and Pre-Copulatory Mate-Guarding in a Fiddler Crab," *Biology Letters* 6, no. 2 (2010): 180–182, doi: 10.1098/rsbl.2009.0767.

Moeliker, C. W. (Kees), "The First Case of Homosexual Necrophilia in the Mallard *Anas platyrhynchos* (Aves: Anatidae)," *Deinsea* 8 (2001): 243–247, www.nmr.nl/nmr/pages/showPage.do?itemid=1930&instanceid=16.

Mollallem, Jon, "Can Animals Be Gay?" *New York Times Magazine*. March 31, 2010, www.nytimes.com/2010/04/04/magazine/04animals-t.html?pagewanted=10.

"Pick Up a Penguin," BBC News, Feb. 26, 1998, http://news.bbc.co.uk/2/hi/asia-pacific/60302.stm.

Tan, M., G. Jones, G. Zhu, J. Ye, T. Hong, et al., "Fellatio by Fruit Bats Prolongs Copulation Time," *PLoS ONE* 4, no. 10 (2009): e7595, doi:10.1371/journal.pone.0007595.

Viegas, Jennifer, "Monkeys, Too, Will Pay for Sex," *Discovery News*, Dec. 19, 2007, http://dsc.discovery.com/news/2007/12/19/monkey-pay-sex.html.

Walker, Matt, "'Sex Pest' Seal Attacks Penguin," BBC News, May 2, 2008, http://news.bbc.co.uk/2/hi/science/nature/7379554.stm.

Ward, Clarissa, "'Panda Porn' to Boost Male's Sex Drive: Scientists Say Showing Sex-Shy Males Videos of Pandas Mating Ramps Up Low Libidos," ABC News, Feb. 15, 2010, http://abcnews.go.com/Nightline/AmazingAnimals/porn-boost-male-pandas-sex-drives/story?id=9718714.

Waterman, J. M., "The Adaptive Function of Masturbation in a Promiscuous African Ground Squirrel," *PLoS ONE* 5, no. 9 (2010): e13060, doi:10.1371/journal.pone.0013060.

Whitfield, John, "Eight-Legged Antics," *Nature News*, July 16, 2001, www.nature.com/news/2001/010716/full/news010719-9.html.

Wolf, Larry, "'Prostitution' Behavior in a Tropical Hummingbird," *Condor* 77 (1975): 140–144.

Yong, Ed, "Squirrels Masturbate to Avoid Sexually Transmitted Infections," *Discover*, Sept. 28, 2010, http://blogs.discovermagazine.com/notrocketscience/2010/09/28/squirrels-masturbate-to-avoid-sexually-transmitted-infections.

4. ANIMAL FAMILY VALUES

Attewill, Fred, "Swan Accused of Killing First Wife, Driving Two More Away," *Metro*, July 29, 2010, www.metro.co.uk/weird/836597-swan-accused-of-killing-first-wife-driving-two-more-away.

Bagemihl, Bruce, *Biological Exuberance: Animal Homosexuality and Natural Diversity*, St. Martin's, 1999, pp. 369, 489, 527, 547, 555, 604, 627–630.

Barash, David, and Judith Eve Lipton, *The Myth of Monogamy: Fidelity and Infidelity in Animals and People*, W. H. Freeman, 2001, p. 47.

Bourton, Jody, "Sexy Sparrow Exposed as World's Most Promiscuous Bird," BBC News, Jan. 22, 2010, http://news.bbc.co.uk/earth/hi/earth_news/newsid_8473000/8473161.stm.

Bradshaw, G. A., *Elephants on the Edge: What Animals Teach Us about Humanity*, Yale University Press, 2010.

Coelho, Sara, "Lesson to Monkey Mothers: Don't Ignore Tantrums," *Science Now*, March 11, 2009, http://news.sciencemag.org/sciencenow/2009/03/11-03.html.

"Diary: New Guinea's Secret Species," BBC News, Feb. 24, 2009, http://news.bbc.co.uk/2/hi/science/nature/7861563.stm.

"Elephants Kill Endangered Rhino," BBC News, Feb. 14, 2000, http://news.bbc.co.uk/2/hi/africa/642731.stm.

Ens, Bruno J., Marcel Kersten, Alex Brenninkmeijer, and Jan B. Hulscher, "Territory Quality, Parental Effort and Reproductive Success of Oystercatchers (*Haematopus ostralegus*)," *Journal of Animal Ecology* 61, no. 3 (1992): 703–715, www.jstor.org/stable/5625.

Frank, L. G., S. E. Glickman, and P. Licht, "Fatal Sibling Aggression, Precocial Development, and Androgens in Neonatal Spotted Hyenas," *Science* 252, no. 5006 (May 3, 1991): 702–704, doi:10.1126/science.2024122.

Fraser, David, and B. K. Thompson, "Armed Sibling Rivalry Among Suckling Piglets," *Behavioral Ecology and Sociobiology* 29 (1991): 9–15, www.springerlink.com/content/j20h64gn84r47q2k.

Gill, Victoria, "Earwigs 'Sniff Out' Best Babies," BBC News, May 13, 2009, http://news.bbc.co.uk/2/hi/science/nature/8040259.stm.

Gilmore, R. G., J. W. Dodrill, and P. A. Linley, "Reproduction and Embryonic Development of the Sand Tiger Shark, *Odontaspis taurus*," *Fishery Bulletin* 81, no. 2 (1983): 201–225.

Hausfater, Glenn, and Sarah Hrdy, eds., *Infanticide: Comparative and Evolutionary Perspectives*, Aldine, 1984, pp. xiv, xxvii, xxvii.

Highfield, Roger, "Elephant Rage: They Never Forgive, Either," *Sydney Morning Herald*, Feb. 17, 2006, www.smh.com.au/articles/2006/02/16/1140064206413.html.

Hrdy, Sarah Blaffer, *Mother Nature: A History of Mothers, Infants, and Natural Selection*, Pantheon, 1999, p. 33, 46, 52.

Ledford, Heidi, " 'Monogamous' Vole in Love-Rat Shock: Randy Rodent Revels in Raunchy Romps," *Nature* 451 (2008): 617, doi:10.1038/451617a.

Leggett, Hadley, "Penguin Parents Won't Chip In to Help Handicapped Spouse," *Wired*, July 2, 2009, www.wired.com/wiredscience/2009/07/penguinpartners.

"Living with Wildlife: Scrubfowl," Northern Territory Government, www.nt.gov.au/nreta/wildlife/animals/living/scrubfowl.html.

Lyon, Bruce E., John M. Eadie, and Linda D. Hamilton, "Parental Choice Selects for Ornamental Plumage in American Coot Chicks," *Nature* 371 (2002): 240–243, www.nature.com/nature/journal/v371/n6494/abs/371240a0.html.

Maestripieri, D., "Early Experience Affects the Intergenerational Transmission of Infant Abuse in Rhesus Monkeys," *Proceedings of the National Academy of Sciences U S A* 102 (2005): 9726–9729, http://primate.uchicago.edu/2005PNAS.pdf.

Maestripieri, D., K. Wallen, and K. A. Carroll, "Infant Abuse Runs in Families of Group-Living Pigtail Macaques," *Child Abuse and Neglect* 21, no. 5 (1997): 465–471, http://primate.uchicago.edu/1997CAN.pdf.

McKie, Robin, "Don't Call Them Love Rats. Virtuous Voles Turn Out to Be All Too Human," *Observer*, Feb. 10, 2008, www.guardian.co.uk/science/2008/feb/10/animalbehaviour.

Mims, Christopher, "Addicted to Love," *Smithsonian Zoogoer* (May–June 2004), http://nationalzoo.si.edu/Publications/ZooGoer/2004/3/monogamy.cfm.

Mock, Douglas, "Infanticide, Siblicide, and Avian Nesting Mortality," in *Infanticide: Comparative and Evolutionary Perspectives*, edited by G. Hausfater and S. G. Hrdy, Aldine, 1984, pp. 3–30.

Norris, Scott, "Murderous Meerkat Moms Contradict Caring Image, Study Finds," *National Geographic News*, March 15, 2006, http://news.nationalgeographic.com/news/2006/03/0315_060315_meerkats.html.

O'Gara, B., "Unique Aspects of Reproduction in the Female Pronghorn (*Antilocapra americana*)," *American Journal of Anatomy* 125 (1969): 217–232.

Ophir, A. G., et al., "Social but Not Genetic Monogamy Is Associated with Greater Breeding Success in Voles," *Animal Behavior* (2008), doi:10.1016/j.anbehav.2007.09.022.

Richardson, Justin, Peter Parnell, and Henry Cole, *And Tango Makes Three*, Simon & Schuster: 2005.

Siebert, Charles, "An Elephant Crackup?" *New York Times Magazine*, Oct. 8, 2006, www.nytimes.com/2006/10/08/magazine/08elephant.html?_r=1&scp=1&sq=an%20elephant%20crackup&st=cse.

Slotow, Rob, Gus van Dyk, Joyce Poole, Bruce Page, and Andre Klocke, "Older Bull Elephants Control Young Males," *Nature* 408 (2000): 425–426, doi:10.1038/35044191.

Sly, Liz, "A Murder Mystery: Why Were Elephants Slaughtering Rhinos?—Lack of Adult Role Models Gets the Blame," *Chicago Tribune/Seattle Times*, Oct. 23, 1994, http://community.seattletimes.nwsource.com/archive/?date=19941023&slug=1937416.

Tucker, Abigail, "The Truth About Lions," *Smithsonian Magazine*, Jan. 2010, www.smithsonianmag.com/science-nature/The-Truth-About-Lions.html.

Wahaj, S. A., N. J. Place, M. L. Weldele, S. E. Glickman, and K. E. Holekamp, "Siblicide in the Spotted Hyena: Analysis with Ultrasonic Examination of Wild and Captive Individuals," *Behavioral Ecology* 18 (2007): 974–984, doi:10.1093/beheco/arm076.

Young, Andrew J., and Tim Clutton-Brock, "Infanticide by Subordinates Influences Reproductive Sharing in Cooperatively Breeding Meerkats." *Biology Letters* 2 (2006): 385–387, http://rsbl.royalsocietypublishing.org/content/2/3/385.full?sid=f2046282-f78b-4c16-8c85-cd225fe35b99.

"Zoo Penguin Couple Breaks Up," *San Francisco Examiner*, July 10, 2009, www.sfexaminer.com/local/Zoo-penguin-couple-breaks-up-50421527.html (accessed July 30, 2009).

Zuk, Marlene, *Sexual Selections: What We Can and Can't Learn About Sex from Animals*, University of California Press, 2003, p. 70.

5. PARTY ANIMALS

"Alcoholic Chimpanzee to Be Taken to Treatment Clinic from Russian Zoo," RIA Novosti, Feb. 26, 2010, http://en.rian.ru/russia/20100226/158020509.html.

"Boozy Bear Plunders Campers' Beer," BBC News, Aug. 19, 2004, http://news.bbc.co.uk/2/hi/3580626.stm.

"Canines Get Caned," *Sydney Morning Herald*, Feb. 18, 2005, www.smh.com.au/articles/2005/02/18/1108609398419.html?oneclick=true.

Charter, David, "Mushrooms Lose Their Magic," *London Times*, Oct. 12, 2007, www.timesonline.co.uk/tol/news/world/europe/article2641369.ece.

Darwin, Charles, *The Descent of Man, and Selection in Relation to Sex*, Clarke, Given and Hooper, 1874.

"Degenerate Honey Bees," *New York Times*, Dec. 26, 1898, http://query.nytimes.com/mem/archive-free/pdf?res=9900E0DD113CE433A25755C2A9649D94699ED7CF.

Dizikes, Cynthia, "Khat—Is It More Coffee or Cocaine?" *Los Angeles Times*, Jan. 3, 2009, http://articles.latimes.com/2009/jan/03/nation/na-khat3.

"Dog Gets High on Pot Found in Park," KING5, Aug. 15, 2009, www.king5.com/news/local/60025647.html.

"Drunk Badger Blocks German Road," AFP, July 8, 2009, www.google.com/hostednews/afp/article/ALeqM5i9a013U8U1EbEBcrlAf0115R6doQ.

"Drunk Elephants Kill Six People," BBC News, Dec. 17, 2002, http://news.bbc.co.uk/2/hi/south_asia/2583891.stm.

Graber, Cynthia, "Fact or Fiction?: Animals Like to Get Drunk," *Scientific American*, July 28, 2008, www.scientificamerican.com/article.cfm?id=animals-like-to-get-drunk.

GrrlScientist, "Lemurs: Millipede Junkies," *Living the Scientific Life*, Jan. 30, 2008, http://scienceblogs.com/grrlscientist/2008/01/lemurs_millipede_junkies.php.

Harrison, David, "Quest for 'Alcohol Gene' Sets Monkeys on Binge," *Telegraph* (UK), March 3, 2002, www.telegraph.co.uk/news/worldnews/centralamericaandthe caribbean/saintkittsandnevis/1386625/Quest-for-alcohol-gene-sets-monkeys-on-binge .html.

Highfield, Roger, "Apes Were First to Get High on Drugs," *Telegraph* (UK), Dec. 26, 2001, www.telegraph.co.uk/news/worldnews/1366327/Apes-were-first-to-get-high -on-drugs.html.

"Honey Bees on Cocaine Dance More, Changing Ideas about the Insect Brain," *Science Daily*, Dec. 25, 2008, www.sciencedaily.com/releases/2008/12/081223091308.htm.

"How Does Catnip Work Its Magic on Cats?" *Scientific American*, May 29, 2007, www .scientificamerican.com/article.cfm?id=experts-how-does-catnip-work-on-cats.

Hussain, Wasbir, "6 Drunk Elephants Electrocute Themselves," AP/*Seattle Times*, Oct. 23, 2007, http://seattletimes.nwsource.com/html/nationworld/2003969241 _webelephants23.html.

"Intoxicated Honey Bees May Clue Scientists into Drunken Human Behavior," *Science Daily*, Oct. 25, 2004, www.sciencedaily.com/releases/2004/10/041025123121.htm.

"Junk Food Turns Rats into Addicts," *Science News*, Nov. 21, 2009, www.sciencenews.org/view/generic/id/48605/title/Junk_food_turns_rats_into_addicts.

Lee, Chris, "Pound for Pound, Bats Can Drink You Under the Table," ARS Technica, Feb. 3, 2010, http://arstechnica.com/science/news/2010/02/not-getting-drunk-has-some -advantages-who-knew.ars.

Mirsch, Laura, "The Dog Who Loved to Suck on Toads," NPR, Oct. 24, 2006, www.npr .org/templates/story/story.php?storyId=6376594.

Moskowitz, Clara, "Tree Shrew Sober Despite Drinking All Day," *Live Science*, July 28, 2008, www.livescience.com/animals/080728-alcoholic-treeshrews.html.

Orbach, D. N., N. Veselka, Y. Dzal, L. Lazure, and M. B. Fenton, "Drinking and Flying: Does Alcohol Consumption Affect the Flight and Echolocation Performance of Phyllostomid Bats?" *PLoS ONE* 5, no. 2 (2010): e8993, doi:10.1371/journal.pone.0008993.

Page, Jeremy, "Villagers Killed as Elephants Develop Taste for Rice Beer," *London Times*, Nov. 15, 2006, www.timesonline.co.uk/tol/news/world/asia/article636819.ece.

Palen, Gary F., and Graham V. Goddard, "Catnip and Oestrous Behaviour in the Cat," *Animal Behaviour* 14, nos. 2–3 (April–July 1966): 372–377, doi:10.1016/S0003 -3472(66)80100-8

Pendergrast, Mark, *Uncommon Grounds: The History of Coffee and How It Transformed Our World*, Basic Books, 2000.

Pfister, J. A., B. L. Stegelmeier, D. R. Gardner, and L. F. James, "Grazing of Spotted Locoweed (*Astragalus lentiginosus*) by Cattle and Horses in Arizona," *Journal of Animal Science* 81 (2003): 2285–2293, www.ars.usda.gov/research/publications/publications. htm?SEQ_NO_115=142862.

Sanders, Laura, "Fruit Flies Can Be Alcoholics Too: Behavior of *Drosophila* Shows Similarities to Human Addiction," *Science News*, Dec. 10, 2009, www.sciencenews.org/view/generic/id/50640/title/Fruit_flies_can_be_alcoholics_too.

Scicurious, "The Mouse That Couldn't Get High," *Neurotopia*, Feb. 4, 2009, http://scienceblogs.com/neurotopia/2009/02/this_is_a_mouse_that_just_cant.php.

Siegel, Ronald, *Intoxication: The Universal Drive for Mind-Altering Substances*, Park Street Press, 2005, pp. 12, 37–39, 65, 119–120, 132, 153–154, 165.

"Stoned Wallabies Make Crop Circles," BBC News, June 25, 2009, http://news.bbc.co.uk/2/hi/asia-pacific/8118257.stm.

"Sweden Woman's 'Murder' Committed by Elk Not Husband," BBC News, Nov. 28, 2009, http://news.bbc.co.uk/2/hi/europe/8384143.stm.

Tewary, Amarnath, "Inside the World's Largest Opium Factory," BBC News, July 21, 2008, http://news.bbc.co.uk/2/hi/south_asia/7509059.stm.

Ustinova, Tanya, "Boozy Chimp Sent to Rehab," Reuters, Feb. 26, 2010, www.reuters.com/article/idUSTRE61P3SM20100226.

Volk, Thomas J., "Tom Volk's Fungus of the Month for December 2000," http://botit.botany.wisc.edu/toms_fungi/dec2000.html.

Walsh, Joseph M., *Coffee: Its History, Classification and Description*, John C. Winston, 1894.

Young, Emma, "Boozing Bees," *New Scientist*, Sept. 26, 2000, p. 38, www.newscientist.com/article/dn22-boozing-bees.html.

6. BEASTLY DEVICES AND DECEITS

Angier, Natalie, "A Highly Evolved Propensity for Deceit," *New York Times*, Dec. 22, 2008, www.nytimes.com/2008/12/23/science/23angi.html?_r=4&8dpc.

Blakeslee, Sandra, "What a Rodent Can Do with a Rake in Its Paw," *New York Times*, March 26, 2008, www.nytimes.com/2008/03/26/science/26rodentw.html.

Brahic, Catherine, "Nut-Cracking Monkeys Find the Right Tool for the Job," *New Scientist*, Jan. 15, 2009, www.newscientist.com/article/dn16426-nutcracking-monkeys-find-the-right-tool-for-the-job.html.

Bro-Jørgensen, Jakob, and Wiline M. Pangle, "Male Topi Antelopes Alarm Snort Deceptively to Retain Females for Mating," *American Naturalist* 176, no. 1 (July 2010), www.journals.uchicago.edu/doi/pdf/10.1086/653078.

Calamia, Joseph, "4 Messages a Pantomiming Orangutan Might Be Trying to Convey," *Discover*, Aug. 11, 2010, http://blogs.discovermagazine.com/80beats/2010/08/11/4-messages-a-pantomiming-orangutan-might-be-trying-to-convey.

Callaway, Ewen, "Pet Dogs Rival Humans for Emotional Satisfaction," *New Scientist*, Jan. 14, 2009, www.newscientist.com/article/dn16412-pet-dogs-rival-humans-for-emotional-satisfaction.html.

Calvin, William H., *Almost Us: Portraits of the Apes*, BookSurge, 2005.

"Cat Parasite Affects Everything We Feel and Do," ABC News, Aug. 9, 2006, http://abcnews.go.com/Technology/DyeHard/story?id=2288095&page=2.

Cheney, D. L., and R. M. Seyfarth, "Truth and Deception in Animal Communication," in *Cognitive Ethology: The Minds of Other Animals*, edited by C. A. Ristau, Lawrence Erlbaum, 1990, pp. 127–152.

Choi, Charles Q., "Chimps Make Spears and Hunt Bushbabies," *Live Science*, Feb. 22, 2007, www.livescience.com/animals/070222_chimp_hunters.html.

Connor, Steve, "Cross-Dressing Cuttlefish Is Casanova of the Reef," *Independent* (UK), Jan. 20, 2005, www.independent.co.uk/environment/crossdressing-cuttlefish-is-casanova-of-the-reef-487413.html.

"Cuttlefish Wimps 'Dress as Girls,'" BBC News, Jan. 19, 2005, http://news.bbc.co.uk/2/hi/science/nature/4188495.stm.

Ehrenberg, Rachel, "Cads of the Savanna: Male Topi Antelopes Lie to Get the Ladies," *Science News* 177, no. 13 (June 19, 2010), www.sciencenews.org/view/generic/id/59497/title/Cads_of_the_savanna.

Ehrlund, Anna, "Never Trust a Hungry Monkey," *Science Now*, June 3, 2009, http://news.sciencemag.org/sciencenow/2009/06/03-02.html.

Flegr, Jaroslav, Jiří Klose Martina Novotná, Miroslava Berenreitterová, and Jan Havlíček, "Increased Incidence of Traffic Accidents in *Toxoplasma*-Infected Military Drivers and Protective Effect RhD Molecule Revealed by a Large-Scale Prospective Cohort Study," *BMC Infectious Diseases* 9 (2009): 72, doi:10.1186/1471-2334-9-72.

Flower, Tom, "Fork-Tailed Drongos Use Deceptive Mimicked Alarm Calls to Steal Food," *Proceedings of the Royal Society Section B*, Nov. 3, 2010, doi:10.1098/rspb.2010.1932.

Geddes, Linda, "Missile-Throwing Chimp Plots Attacks on Tourists," *New Scientist*, March 9, 2009, www.newscientist.com/article/dn16726-missilethrowing-chimp-plots-attacks-on-tourists.html.

Hanlon, Roger T., Marié-Jose Naud, Paul W. Shaw, and Jon N. Havenhand, "Transient Sexual Mimicry Leads to Fertilization," *Nature* 433 (2005): 212, doi:10.1038/433212a.

Hauser, Marc D., *Wild Minds: What Animals Really Think*, Holt Paperbacks, 2001, pp. 142, 147, 149, 161.

Holland, Jennifer S., "Let Sneaky Dogs Lie," NGM Blog Central, Aug. 13, 2010, http://blogs.ngm.com/blog_central/2010/08/let-sneaky-dogs-lie.html.

"House Cats Know What They Want and How to Get It from You," *Science Daily*, July 13, 2009, www.sciencedaily.com/releases/2009/07/090713121348.htm.

Hrdy, Sarah Blaffer, *Mother Nature: A History of Mothers, Infants, and Natural Selection*, Pantheon, 1999, p. 75.

Kennedy, J. Michael, "Octavia the Octopus Dies as Tank Empties," *Los Angeles Times*, April 12, 1994, http://articles.latimes.com/1994-04-12/local/me-44986_1_san-pedro-aquarium.

Kundey, Shannon M. A., Andres de Los Reyes, Chelsea Taglang, Rebecca Allen, Sabrina Molina, Erica Royer, and Rebecca German, "Domesticated Dogs (*Canis familiaris*) React to What Others Can and Cannot Hear." *Applied Animal Behaviour Science* 126, no. 1 (Aug. 2010): 45–50, doi:10.1016/j.applanim.2010.06.002.

"Ladyboy Lizards Use Transvestite Trickery," AFP, March 2, 2009, www.google.com/hostednews/afp/article/ALeqM5h5PZuDpvK4jur27xYvDM4AGRT_MQ.

Leake, Jonathan, "Dangerrrr: Cats Could Alter Your Personality," *LondonTimes*, Sept. 21, 2003, www.timesonline.co.uk/tol/news/uk/health/article1161725.ece.

Linden, Eugene, *The Octopus and the Orangutan; More True Tales of Animal Intrigue, Intelligence, and Ingenuity*, Plume, 2003, p. 10.

———, *The Parrot's Lament, and Other True Tales of Animal Intrigue, Intelligence, and Ingenuity*, Dutton, 1999, p. 65, 145, 148.

Lombardi, Linda, "Fireflies Shine Light on Insect Conservation," AP, June 10, 2009. www.lindalombardi.com/storage/fireflies.pdf.

Marshall, Michael, "Dogs Know That Stealth Pays When Your Eyes Are Averted," *New Scientist*, July 24, 2010, www.newscientist.com/article/mg20727703.400-sneaky-dogs-take-food-quietly-to-avoid-getting-caught.html.

McVeigh, Karen, "Sound Effect: How Cats Exploit the Human Need to Nurture," *Guardian* (UK), July 14, 2009, www.guardian.co.uk/science/2009/jul/13/cats-purr-food-research.

Miyatake, Takahisa, Satoshi Nakayama, Yusuke Nishi, and Shuhei Nakajima, "Tonically Immobilized Selfish Prey Can Survive by Sacrificing Others," *Proceedings of the Royal Society Section B*, April 29, 2009, doi:10.1098/rspb.2009.0558.

Møller, Anders P., "Deceptive Use of Alarm Calls by Male Swallows, *Hirundo rustica*: A New Paternity Guard," *Behavioral Ecology* 1, no. 1 (1990): 1–6, doi:10.1093/beheco/1.1.1.

———, "False Alarm Calls as a Means of Resource Usurpation in the Great Tit *Parus major*," *Ethology* 79 (1988): 25–30, doi:10.1111/j.1439-0310.1988.tb00697.x.

Nagasawa, Miho, Kazutaka Mogi, and Takefumi Kikusui, "Attachment Between Humans and Dogs," *Japanese Psychological Research* 51, no. 3 (2009): 209–221, http://onlinelibrary.wiley.com/doi/10.1111/j.1468-5884.2009.00402.x/pdf.

Okanoya, K., N. Tokimoto, N. Kumazawa, S. Hihara, and A. Iriki, "Tool-Use Training in a Species of Rodent: The Emergence of an Optimal Motor Strategy and Functional Understanding," *PLoS ONE* 3, no. 3 (2008): e1860, doi:10.1371/journal.pone.0001860.

"Orangutan Makes Break for Freedom at Audubon Zoo," WWL, Jan. 30, 2009, www.wwl.com/Orangutan-makes-break-for-freedom-at-Audubon-Zoo/3759550 (accessed Feb. 7, 2009).

"Orang-utan Short-Circuits Electric Fence in Zoo 'Escape,'" *Telegraph* (UK), May 10, 2009, www.telegraph.co.uk/news/newstopics/howaboutthat/5304626/Orang-utan-short-circuits-electric-fence-in-Zoo-escape.html.

Osvath, Mathias, "Spontaneous Planning for Future Stone Throwing by a Male Chimpanzee," *Current Biology* 19, no. 5 (2009): R190–R191, www.cell.com/current-biology/fulltext/S0960-9822%2809%2900547-8.

"Otto the Octopus Wreaks Havoc," *Telegraph* (UK), Oct. 31, 2008, www.telegraph.co.uk/news/newstopics/howaboutthat/3328480/Otto-the-octopus-wrecks-havoc.html.

Owen, Michael, "Grief at Losing Partner Sends Karta Over Zoo Wall," *Australian*, May 11, 2009, www.theaustralian.com.au/news/grief-sends-karta-over-zoo-wall/story-e6frg6no-1225710738150.

Pickrell, John, "Crows Better at Tool Building Than Chimps, Study Says," *National Geographic News*, April 23, 2003, http://news.nationalgeographic.com/news/2003/04/0423_030423_crowtools.html.

Pool, Bob, "Octopus Floods Santa Monica Pier Aquarium," *Los Angeles Times*, Feb. 27, 2009, http://articles.latimes.com/2009/feb/27/local/me-octopus27.

Pope, John, "Escape Artist: An Audubon Zoo Orangutan Boldly Goes Where No Orangutan Has Gone Before—Over the Wall," NOLA, Jan. 30, 2009, www.nola.com/news/index.ssf/2009/01/ape_escape_audubon_zoo_orangut.html.

Roylance, Frank D., "Researchers Explore Link Between Schizophrenia, Cat Parasite," Physorg, Aug. 4, 2010, www.physorg.com/news200164470.html.

Russon, Anne, and Kristin Andrews, "Orangutan Pantomime: Elaborating the Message," *Biology Letters*, Aug. 11, 2010, doi:10.1098/rsbl.2010.0564.

Sample, Ian, "Chimp Who Threw Stones at Zoo Visitors Showed Human Trait, Says Scientist," *Guardian* (UK), March 9, 2009, www.guardian.co.uk/science/2009/mar/09/chimp-zoo-stones-science.

———, "Orangutans Use Mime to Make Themselves Understood," *Guardian* (UK), Aug. 11, 2010, www.guardian.co.uk/science/2010/aug/11/orangutans-mime.

Santos, Laurie R., Aaron G. Nissen, and Jonathan A. Ferrugia, "Rhesus Monkeys, *Macaca mulatta*, Know What Others Can and Cannot Hear," *Animal Behaviour* 71, no. 5 (May 2006): 1175–1181, www.yale.edu/monkeylab/Main/Publications_files/sdarticle%283%29.pdf.

Scheldeman, Griet, "Orangutan Ruse Misleads Predators," BBC News, Aug. 5, 2009, http://news.bbc.co.uk/2/hi/science/nature/8184015.stm.

Skloot, Rebecca, "'Cat Lady' Conundrum," *New York Times*, Dec. 9, 2007, www.nytimes.com/2007/12/09/magazine/09_10_catcoat.html.

"The Story of an Octopus Named Otto," NPR, Nov. 2, 2008, www.npr.org/templates/story/story.php?storyId=96476905.

Trivedi, Bijal, "Chimps Shown Using Not Just a Tool but a 'Tool Kit,'" *National Geographic News*, Oct. 6, 2004, http://news.nationalgeographic.com/news/2004/10/1006_041006_chimps.html.

Viegas, Jennifer, "Orangutans Use Charade-Like Communication," *Discovery News*, Aug. 10, 2010, http://news.discovery.com/animals/orangutans-pantomime-charade.html.

Waal, Frans de, *Chimpanzee Politics*, Johns Hopkins University Press, 1998.

Walker, Matt, "Chimps Use Cleavers and Anvils as Tools to Chop Food," BBC News, Dec. 24, 2009, http://news.bbc.co.uk/earth/hi/earth_news/newsid_8427000/8427974.stm.

Wheeler, Brandon C., "Monkeys Crying Wolf? Tufted Capuchin Monkeys Use Anti-Predator Calls to Usurp Resources from Conspecifics," *Proceedings of the Royal Society Section B*, June 3, 2009, pp. 3013–3018, doi:10.1098/rspb.2009.0544.

Whiting, Martin J., Jonathan K. Webb, and J. Scott Keogh, "Flat Lizard Female Mimics Use Sexual Deception in Visual but Not Chemical Signals," *Proceedings of the Royal Society Section B*, Feb. 25, 2009, doi:10.1098/rspb.2008.1822.

Winkler, Robert, "Crow Makes Wire Hook to Get Food," *National Geographic News*, Aug. 8, 2002, http://news.nationalgeographic.com/news/2002/08/0808_020808_crow.html.

Yong, Ed, "The Bird That Cries Hawk: Fork-Tailed Drongos Rob Meerkats with False Alarms," *Discover*, Nov. 3, 2010, http://blogs.discovermagazine.com/notrocketscience/2010/11/03/the-bird-that-cries-hawk-fork-tailed-drongos-rob-meerkats-with-false-alarms.

7. MASTERS OF MISDIRECTION

"Amorous Dolphin Targeting Swimmers," CNN, June 4, 2002, http://archives.cnn.com/2002/WORLD/europe/06/04/uk.dolphin/index.html.

Angier, Natalie, "Dolphin Courtship: Brutal, Cunning and Complex," *New York Times*, Feb. 18, 1992, www.nytimes.com/1992/02/18/science/dolphin-courtship-brutal-cunning-and-complex.html.

Audi, Tamara, "It's Easter. Where's Elmer Fudd When You Need Him?" *Wall Street Journal*, April 3, 2010, http://online.wsj.com/article/SB10001424052702303395904575158581521399358.html.

Broad, William J., "Evidence Puts Dolphins in New Light, As Killers," *New York Times*, July 6, 1999, www.nytimes.com/1999/07/06/health/evidence-puts-dolphins-in-new-light-as-killers.html.

Carlson, Kathryn Blaze, "Victoria Rabbits Hop for the Border," *National Post*, Sept. 14, 2010, www.vancouversun.com/technology/Victoria+rabbits+border/3520105/story.html (accessed Sept. 20, 2010).

Chudler, Eric H., "Brain Basics: How Much Do Animals Sleep?" Neuroscience for Kids, http://faculty.washington.edu/chudler/chasleep.html.

Coniff, Richard, *Swimming with Piranhas at Feeding Time: My Life Doing Stupid Things with Animals*. Norton, 2009.

Connor, Steve, "Jurassic Bully: Pick on Someone Your Own Size . . ." *Independent* (UK), Aug. 5, 2009, www.independent.co.uk/news/science/jurassic-bully-pick-on-someone-your-own-size-1767370.html.

"The Dangers of Dolphins," BBC News, Sept. 11, 2002, www.bbc.co.uk/southampton/features/dolphins/george.shtml.

Dunn, Dale G., Susan G. Barco, D. Ann Pabst, and William A. McLellan, "Evidence for Infanticide in Bottlenose Dolphins of the Western North Atlantic," *Journal of Wildlife Diseases* 38, no. 3 (2002): 505–510.

"European Wild Rabbit (*Oryctolagus cuniculus*)," Australian Department of Environment and Heritage, 2004, www.environment.gov.au/biodiversity/invasive/publications/pubs/rabbit.pdf.

Foundation for Rabbit-Free Australia Inc., www.rabbitfreeaustralia.org.au.

Hooper, Rowan, "Spear-Wielding Chimps Snack on Skewered Bushbabies," *New Scientist*, Feb. 22, 2007, www.newscientist.com/article/dn11234.

"Hummingbird Behavior," World of Hummingbirds, www.worldofhummingbirds.com/behavior.php.

"The Impacts of Bycatch," Environmental Justice Foundation, www.ejfoundation.org/page173.html.

"Judge Strikes Down Injunction, Allows University of Victoria to Trap Rabbits," *Vancouver Sun*, Aug. 30, 2010, www.vancouversun.com/technology/Judge+strikes+down+injunction+allows+University+Victoria+trap+rabbits/3462548/story.html (accessed Sept. 20, 2010).

Kaplan, Matt, "'Loving' Bonobos Seen Killing, Eating Other Primates," *National Geographic News*, Oct. 13, 2008, http://news.nationalgeographic.com/news/2008/10/081013-bonobos-attack-missions_2.html.

Leake, Jonathan, "Scientists Say Dolphins Should Be Treated as 'Non-Human Persons,'" *Times*, Jan. 3, 2010, www.timesonline.co.uk/tol/news/science/article6973994.ece.

"LMU Professor Presents Case for Dolphins as Nonhuman Persons at Science Conference," Loyola Marymount University (press release), Feb. 22, 2010, http://newsroom.lmu.edu/newsroompressreleases/dolphins.htm.

McCleod, Mairi, "'Hippy' Monkey Is a Killer When Starved of Sex," *New Scientist*, July 4, 2009, www.newscientist.com/article/mg20327154.300-hippy-monkey-is-a-killer-when-starved-of-sex.html.

Milius, S., "Infanticide Reported in Dolphins," *Science News* 154, no. 3 (1998): p. 36, www.sciencenews.org/sn_arc98/7_18_98/fob1.htm.

Newby, Jonica, "Dolphin Talk," Australian Broadcasting Corporation, Feb. 26, 2004, www.abc.net.au/catalyst/stories/s1052990.htm.

Norris, S., M. Hall, E. Melvin, and J. Parrish, "Thinking Like an Ocean: Ecological Lessons from Marine Bycatch," *Conservation Magazine* 3, no. 4 (2002): 10–19, www.conservationmagazine.org/2008/07/thinking-like-an-ocean.

Patterson, I. A. P., R. J. Reid, B. Wilson, K. Grellier, H. M. Ross, and P. M. Thompson, "Evidence for Infanticide in Bottlenose Dolphins: An Explanation for Violent Interactions with Harbour Porpoises?" *Proceedings. Biological Sciences/The Royal Society* 265, no. 1402 (1998): 1167–1170.

Pryor, Karen, "Why Porpoise Trainers Are Not Dolphin Lovers: Real and False Communication in the Operant Setting," *Annals of the New York Academy of Sciences* 364 (1981): 137–143, doi:10.1111/j.1749-6632.1981.tb34467.x.

Shiffman, David (WhySharksMatter), "Dolphin-Safe Tuna: Conservation Success Story or Ecological Disaster?" *Southern Fried Science*, July 26, 2010, www.southernfriedscience.com/?p=6539.

Surbeck, Martin, and Gottfried Hohmann, "Primate Hunting by Bonobos at LuiKotale, Salonga National Park," *Current Biology* 18, no. 19 (2008): R906–R907.

Talebi, M. G., R. Beltrão-Mendes, and P. C. Lee, "Intra-Community Coalitionary Lethal Attack of an Adult Male Southern Muriqui (*Brachyteles arachnoids*)," *American Journal of Primatology* 71, no. 10 (2009): 860–867.

Tickle, Louise, "Dolphins Bully Porpoises, Researcher Discovers," *Guardian* (UK), May 11, 2010, www.guardian.co.uk/education/2010/may/11/dolphin-research-bangor.

Tucker, Abigail, "The Truth about Lions," *Smithsonian*, Jan. 2010, www.smithsonianmag.com/science-nature/The-Truth-About-Lions.html.

White, Thomas I., *In Defense of Dolphins: The New Moral Frontier*, Wiley-Blackwell, 2007.

Wrangham, Richard W., and Dale Peterson, *Demonic Males: Apes and the Origins of Human Violence*, Houghton Mifflin Harcourt, 1997.

Yong, Ed, "Chimpanzees Murder for Land," *Discover*, June 21, 2010, http://blogs.discovermagazine.com/notrocketscience/2010/06/21/chimpanzees-murder-for-land.

8. WITH FRIENDS LIKE THESE . . .

Anderson, Joel, "Hernando Man Run Over by Own Truck after His Dog Puts It in Gear," *St. Petersburg Times*, June 23, 2010, www.tampabay.com/news/publicsafety/accidents/hernando-man-run-over-by-own-truck-after-his-dog-puts-it-in-gear/1104156.

Bradley, Janis, *Dogs Bite, but Balloons and Slippers Are More Dangerous*, James and Kenneth Publishers, 2005.

Bryner, Jeanna, "Cats and Dogs Are Household Hazards," *Live Science*, April 16, 2010, www.livescience.com/animals/tripping-over-cats-dogs-100416.html.

Cerve, Kate, "Dog Chows Down on SC Man's School Board Petition," *Rock Hill (SC) Herald*, Aug. 13, 2010, www.heraldonline.com/2010/08/13/2380523/dog-chows-down-on-sc-mans-school.html (accessed Aug. 20, 2010).

Chang, Sophia, "Dog Rolls Van Through St. James Coffee Shop Window," *Newsday*, Nov. 20, 2008, www.newsday.com/long-island/dog-rolls-van-through-st-james-coffee-shop-window-1.884845.

Coles, John, "Posties Dodge 'Menace' Moggy," *Sun* (UK), Oct. 8, 2009, www.thesun.co.uk/sol/homepage/news/2673393/Posties-dodge-menace-moggy.html.

Coppinger, Raymond, and Lorna Coppinger, *Dogs: A New Understanding of Canine Origin, Behavior, and Evolution*, University of Chicago Press, 2002.

Coppinger, Ray, and Richard Schneider, "Evolution of Working Dogs," in *The Domestic Dog: Its Evolution, Behaviour, and Interactions with People*, edited by James Serpell, Cambridge University Press, 1995. 21–47.

Cuttle, Vivien, "Pets Pose Bigger Risk Than Wildlife," Australian Broadcasting Corporation, April 8, 2010, www.abc.net.au/news/stories/2010/04/08/2867852.htm.

"Dog Drives Owner's Car into River," MSNBC, June 22, 2007, www.msnbc.msn.com/id/19380779.

"Dog Eats $400, but Woman Recovers Some of It," AP/MSNBC, March 18, 2009, www.msnbc.msn.com/id/29760059/22027217.

"Dogs and More Dogs," *Nova*, Feb. 3, 2004, www.pbs.org/wgbh/nova/transcripts/3103_dogs.html.

"Dogs Likely Descended from Middle Eastern Wolf," NPR, March 18, 2010, www.npr.org/templates/transcript/transcript.php?storyId=124768140.

"Dog Swallows $20,000 diamond in Maryland Jewelry Store," WJLA, March 11, 2010, www.wjla.com/news/stories/0310/715111.html.

"Driving Dog Let Off with Warning," TVNZ, Sept. 16, 2009, http://tvnz.co.nz/national-news/driving-dog-let-off-warning-after-crash-2992911.

Duffy, Deborah L., Yuying Hsu, and James A. Serpell, "Breed Differences in Canine Aggression," *Applied Animal Behaviour Science* 114, no. 3 (2008): 441–460, www.journals.elsevierhealth.com/periodicals/applan/article/PIIS0168159108001147/abstract.

Hare, Brian, Michelle Brown, Christina Williamson, and Michael Tomasello, "The Domestication of Social Cognition in Dogs," *Science* 298, no. 5598 (2002): 1634–1636, doi:10.1126/science.1072702.

Kennedy, William, "A Mighty Heart," PEN American Center, www.pen.org/viewmedia.php/prmMID/1101/prmID/510.

Lite, Jordan, "Pet Dogs and Cats a Good Way to Break a Leg, Government Says," *Scientific American*, March 26, 2009, www.scientificamerican.com/blog/post.cfm?id=pet-dogs-and-cats-a-good-way-to-bre-2009-03-26.

"Man on Hunt Trip Shot by His Dog," *Sun* (UK), July 13, 2010, www.thesun.co.uk/sol/homepage/news/3051943/Man-on-hunt-trip-shot-by-his-DOG.html.

"Nonfatal Fall-Related Injuries Associated with Dogs and Cats—United States, 2001–2006," *Morbidity and Mortality Weekly Report*, Centers for Disease Control and Prevention 58, no. 11 (March 27, 2009): 277–281, www.cdc.gov/mmwr/preview/mmwrhtml/mm5811a1.htm.

Pate, Karen, "Tillamook-Area Man Shot by Dog Recovering," *Oregonian*, Nov. 23, 2008, www.oregonlive.com/news/index.ssf/2008/11/tillamookarea_man_shot_by_dog.html.

"Post Ban Threat over Aggressive Cat," *Sky News*, Jan. 16, 2007, http://news.sky.com/skynews/Home/Post-Ban-Threat-Over-Aggressive-Cat/Article/20070131247554.

"Recurring Themes: Dogs Causing Trouble," Chuck Shepherd's News of the Weird, March 1, 2009, www.newsoftheweird.com/archive/nw090301.html.

Rosenkrans, Nolan, "Dog Vomit Leads to Car Crash," *Winona Daily News*, April 27, 2010, www.winonadailynews.com/news/local/crime-and-courts/article_4d9d90ba-51b5-11df-a8d3-001cc4c002e0.html.

"Royal Mail Refuses to Post Family's Letters Due to Household Cat," *Telegraph* (UK), Nov. 1, 2010, www.telegraph.co.uk/news/newstopics/howaboutthat/8102124/Royal-Mail-refuses-to-post-familys-letters-due-to-household-cat.html.

"Royal Mail Threatens Delivery Withdrawal After Postman 'Attacked' by Kitten," *Telegraph* (UK), June 20, 2009, www.telegraph.co.uk/news/newstopics/howaboutthat/5582674/Royal-Mail-threatens-delivery-withdrawal-after-postman-attacked-by-kitten.html.

Savill, Richard, "Postman Pat and the Psychopathic Cat," *Telegraph* (UK), Aug. 23, 2002, www.telegraph.co.uk/news/uknews/1405165/Postman-Pat-and-the-psychopathic-cat. . . .html.

Sims, Paul, "Warning: Dangerous Cat—Has Attacked 13 People in the Last Six Years," *Daily Mail*, Dec. 26, 2006, www.dailymail.co.uk/news/article-424841/Warning-Dangerous-Cat—attacked-13-people-years.html.

Thorndike, Edward, *Animal Intelligence*, Macmillan, 1911.

"Tiger the Cat Puts a Stop to Royal Mail Deliveries," *Telegraph* (UK), April 9, 2010, www.telegraph.co.uk/news/newstopics/howaboutthat/7568434/Tiger-the-cat-puts-a-stop-to-Royal-Mail-deliveries.html.

"True Stories of Incredible Pet Mishaps," VPI Hambone Award, www.vpihamboneaward.com.

"Violent Cat Leaves Postmen Running Scared," *Telegraph* (UK), Dec. 24, 2007, www.telegraph.co.uk/news/uknews/1573547/Violent-cat-leaves-postmen-running-scared.html.

Wardrop, Murray, "Police Officer Attacked by Her Own Dog After Being Shot by Robber," *Telegraph* (UK), April 30, 2009, www.telegraph.co.uk/news/uknews/law-and-order/5253428/Police-officer-attacked-by-her-own-dog-after-being-shot-by-robber.html.

Wilson, Nick, "Man Claims Dog Tripped Him, Caused Him to Fatally Shoot Wife," *San Luis Obispo Tribune*, Nov. 18, 2009, www.sanluisobispo.com/2009/11/17/925151/man-claims-dog-tripped-him-caused.html.

———, "San Miguel Man Won't Be Retried for Death of Wife," *San Luis Obispo Tribune*, July 29, 2010, www.sanluisobispo.com/2010/07/29/1231668/san-miguel-man-wont-be-retried.html.

"Woman Charged with Fatally Shooting Husband after Blaming Dog," WMBB, Nov. 14, 2009, www.panhandleparade.com/index.php/mbb/article/bcso_investigates_fatal_shooting/mbb7719882.

"Wombat Mauls Bushfire Survivor," Australian Broadcasting Corporation, April 6, 2010, www.abc.net.au/news/stories/2010/04/06/2865005.htm.

Woodward, Sommer, "Driving Dog Has Accident at Car Wash," *Daily Times*, Nov. 6, 2008, http://pryordailytimes.com/local/x519317861/Driving-dog-has-accident-at-car-wash.

9. OUR OWN WORST ENEMIES

"Baboons Go Ape over S. Africa's Wine Crop," MSNBC, March 23, 2010, www.msnbc.msn.com/id/36005735/ns/world_news-africa.

"Bond with Dog Helped Girl Survive: Dog 'Blue' Saved 3-Year-Old Victoria Bensch's Life, Authorities Say," KPHO, Feb. 19, 2010, www.kpho.com/news/22609878/detail.html.

"Chesapeake Veteran Says Attack by His Service Monkey Was Worse Than Combat," WTKR, www.wtkr.com/news/wtkr-ches-veteran-monkey-attack,0,2639004.story.

Drummond, Andrew, "British Woman Trying to Overcome Life-Long Fear of Primates Is Attacked by Macaques in Thailand," *Daily Mail*, June 18, 2010, www.dailymail.co.uk/

news/worldnews/article-1287381/British-woman-trying-overcome-life-long-fear-primates-attacked-crab-eating-Macaques-Thailand.html.

Fahrenthold, David A., "Survivors of Attacks Sink Teeth into Fight to Save Sharks," *Washington Post*, July 15, 2009, www.washingtonpost.com/wp-dyn/content/article/2009/07/14/AR2009071403248.html.

Heilprin, John, "Shark Attack Survivors Don't Forget but Do Forgive,"AP/*Boston Herald*, Sept. 13, 2010, www.bostonherald.com/news/national/general/view/20100913shark_attack_survivors_dont_forget_but_do_forgive.

Hinckley, David, "Tammy the Turnpike Turkey Plays Chicken with New Jersey Turnpike Tollbooth . . . and Loses," *New York Daily News*, Nov. 18, 2009, www.nydailynews.com/lifestyle/pets/2009/11/18/2009-11-18_wild_turkey.html#ixzz0XKaxQPQ7.

Jiwa, Salim, "Stray Dog Kept Missing Tot Company in Yukon Wilds," *Vancouverite*, Sept. 4, 2009, www.vancouverite.com/2009/09/04/stray-dog-kept-missing-tot-company-in-yukon-wilds.

Kettler, Shannon, "Ky. Man Gets Up Close with an Ape at Zoo," *Kentucky Post*, Aug. 22, 2008, www2.kypost.com/dpp/news/local_news/Ky.-Man-Gets-Up-Close-With-An-Ape-At-Zoo.

Leake, Jonathan, "Yankee Tree Rats, You're Brits Now," *Times*, April 4, 2010, www.timesonline.co.uk/tol/news/science/living/article7086785.ece.

"Lost Child Discovered Nearby with Loyal Dog," *Ocala Star-Banner*, Feb. 3, 1984, p. 1B, http://news.google.com/newspapers?nid=hXZnTIgIr50C&dat=19840203&printsec=frontpage.

Malkin, Bonnie, "Man in Hospital after Wombat Attack," *Telegraph* (UK), April 6, 2010, www.telegraph.co.uk/news/7558903/Man-in-hospital-after-wombat-attack.html.

"Man Bites Panda After Zoo Attack," BBC News, Sept. 20, 2006, http://news.bbc.co.uk/2/hi/asia-pacific/5364058.stm.

"Monkey Again Bites Owner in Chesapeake," WVEC, March 30, 2010, www.wvec.com/news/local/Monkey-again-bites-owner-in-Chesapeake-89485767.html.

"Panda Bites Student Who Wanted a Hug," *China Daily*, Nov. 23, 2008, www.chinadaily.com.cn/china/2008-11/23/content_7231170.htm.

Pulliam, Christine, "Astronomers Solve Mystery of Dusty Foot Trails Crossing Telescope Mirrors," Smithsonian Science, Sept. 9, 2009, http://smithsonianscience.org/2009/09/smithsonian-astronomers-in-arizona-identify-mystery-nocturnal-visitor.

"Two Fined in Lincoln Park Zoo Bear Attack," WTAQ, March 11, 2010, www.wtaq.com/news/articles/2010/mar/11/two-fined-lincoln-park-zoo-bear-attack.

"Woman Enters Exhibit, Elephant Smacks Her," AP/WJZ, March 4, 2006, http://wjz.com/watercooler/Water.Cooler.Waco.2.264774.html (accessed Feb. 15, 2010).

"Zoo Meerkats Killed, Test Negative for Rabies," WCCO, Aug. 4, 2006, http://wcco.com/topstories/meerkats.Minnesota.Zoo.2.360637.html (accessed Feb. 15, 2010).

"Zoo Mulls Next Step in Wolf Case," *Chicago Tribune*, Nov. 20, 2003, http://pqasb.pqarchiver.com/chicagotribune/access/454879121.html?FMT=ABS&FMTS=

ABS:FT&date=Nov+20%2C+2003&author=&pub=Chicago+Tribune&edition=&startpage=3&desc=Zoo+mulls+next+step+in+wolf+case.

"Zoo's Penguin Afraid of Water," UPI, Jan. 22, 2009, www.upi.com/Odd_News/2009/01/22/Zoos-penguin-afraid-of-water/UPI-33331232667971.

10. UNGRATEFUL BEASTS

"The Amazing Monkey Waiters That Serve Tables in a Japanese Restaurant," *Daily Mail*, Oct. 7, 2008, www.dailymail.co.uk/news/worldnews/article-1071289/Pictured-The-amazing-monkey-waiters-serve-tables-Japanese-restaurant.html.

The Asian Elephant Art and Conservation Project, www.elephantart.com.

"BBC Cameraman Attacked by Monkey at Commonwealth Games," BBC News, Oct. 6, 2010, http://news.bbc.co.uk/local/norfolk/hi/people_and_places/newsid_9068000/9068124.stm.

Blanchard, Ben, "Grumpy Miracle Pig Voted Most Popular Animal," Reuters, Dec. 22, 2008, www.reuters.com/article/idUSTRE4BL31A20081222.

Bonnett, Tom, "Thai Police Force Takes Monkey on the Beat," *Sky News*, April 2, 2010, http://news.sky.com/skynews/Home/Strange-News/Monkey-Police-Macaque-On-Patrol-With-Thai-Force-To-Help-Improve-Relations-With-Muslim-Separatists/Article/201004115591175?.

"Bridges Help Dormice to Cross Church Village Bypass," BBC News, Aug. 25, 2010, www.bbc.co.uk/news/uk-wales-south-east-wales-11082007.

"The Brown Treesnake on Guam," U.S. Geological Survey, May 23, 2008, www.fort.usgs.gov/resources/education/bts.

Carey, Elizabeth, "Amazing Dog Rescue on Lake Erie," *Buffalo Business First*, March 16, 2010, http://buffalo.bizjournals.com/buffalo/blog/stay_tuned/2010/03/amazing_dog_rescue_on_lake_erie.html.

"Count Your Chickens," Darwin Awards, 1995, www.darwinawards.com/darwin/darwin1995-01.html.

Demetriou, Danielle, "Two Monkeys Appointed Station Masters at Japanese Train Station," *Telegraph* (UK), Oct. 18, 2010, www.telegraph.co.uk/news/newstopics/howaboutthat/8070139/Two-monkeys-appointed-station-masters-at-Japanese-train-station.html.

Discepolo, John, "Wild Turkey Rules the Roost in North Bend Neighborhood," KOMO News, May 5, 2010, www.komonews.com/news/local/92920119.html.

"Dog Rescued from Ice Twice," UPI, March 17, 2010, www.upi.com/Odd_News/2010/03/17/Dog-rescued-from-ice-twice/UPI-95001268856214.

"Dog Saved from Flooded L.A. River Is Safe; Firefighter Bitten During Rescue Released from Hospital," *Los Angeles Times*, Jan. 22, 2010, http://latimesblogs.latimes.com/unleashed/2010/01/dog-saved-from-flooded-la-river-is-safe-firefighter-bitten-during-rescue-released-from-hospital-.html.

"Electrical Problems Caused by the Brown Treesnake," U.S. Geological Survey, July 26, 2005, www.fort.usgs.gov/resources/education/bts/impacts/electrical.asp.

"Gay Penguin Pair Tie the Knot," *Sun* (UK), Jan. 27, 2009, www.thesun.co.uk/sol/homepage/news/article2176812.ece.

Gefter, Amanda, "Film Festival: 'A Well-Trained Monkey Could Do My Job,'" *New Scientist*, Oct. 22, 2010, www.newscientist.com/blogs/culturelab/2010/10/film-festival-a-well-trained-monkey-could-do-my-job.html.

———, "Film Festival: Can Monkeys Make Movies?" *New Scientist*, Oct. 15, 2010, www.newscientist.com/blogs/culturelab/2010/10/film-festival-can-monkeys-make-films.html.

"Grizzly Bear Is Best Man at Wedding," *Telegraph* (UK), April 29, 2009, www.telegraph.co.uk/news/newstopics/howaboutthat/5240623/Grizzly-bear-is-best-man-at-wedding.html.

"Highway Man Settles In," *San Jose Mercury News*, June 1, 2010, www.mercurynews.com/ci_15179342 (accessed Feb. 15, 2010).

"History of the Brown Treesnake Invasion on Guam," U.S. Geological Survey, July 26, 2005, www.fort.usgs.gov/resources/education/bts/invasion/history.asp.

"Italian Villages Terrorised by Rampaging Bear but Law Protects Animal," *Telegraph* (UK), May 21, 2010, www.telegraph.co.uk/news/worldnews/europe/italy/7750045/Italian-villages-terrorised-by-rampaging-bear-but-law-protects-animal.html.

Janus, Allan, *Animals Aloft: Photographs from the Smithsonian National Air and Space Museum*, Bunker Hill Publishing, 2005.

Javed, Noor, "Can a Dog Receive Communion?" *Toronto Star*, July 22, 2010, www.thestar.com/news/gta/article/838717--can-a-dog-receive-communion.

"Jumping for Joy: Australian Couple Have Kangaroo as Bridesmaid on Their Wedding Day," *Daily Mail*, May 11, 2010, www.dailymail.co.uk/news/worldnews/article-1276183/Jumping-joy-Australian-couple-kangaroo-bridesmaid-wedding-day.html.

Killen, John, "Oregon Sea Lions Making Their Mark in the Art World," *Oregonian*, Aug. 30, 2009, www.oregonlive.com/news/index.ssf/2009/08/newport_when_marine_mammalogis.html.

Langdon, Sarah, "Nonja, the Orangutan Photographer," *National Geographic*, Dec. 7, 2009, http://blogs.nationalgeographic.com/blogs/intelligenttravel/2009/12/nonja-the-orangutan-photograph.html.

Lawrence, Edwin, "Runaway Capybara Is Back at Heads of Ayr Farm Park," *Ayrshire Post*, Jan. 8, 2010, www.ayrshirepost.net/ayrshire-news/local-news-ayrshire/ayr-news/2010/01/08/runaway-capybara-is-back-at-heads-of-ayr-farm-park-102545-25538073.

Markey, Sean, "Giant Mice Devouring Island Seabird Chicks, Threatening Extinction," *National Geographic News*, April 13, 2007, http://news.nationalgeographic.com/news/2007/04/070413-mice-birds.html.

"Never Slaughter a Chicken in Front of a Monkey," *Orange News* (UK), July 5, 2010, http://web.orange.co.uk/article/quirkies/Never_slaughter_a_chicken_in_front_of_a_monkey.

"No, I'm Not Going to Swan Off! Bird Causes Traffic Chaos in London After Resisting Multiple Rescue Attempts," *Daily Mail*, March 9, 2010, www.dailymail.co.uk/news/

article-1256463/No-Im-going-swan-Bird-causes-traffic-chaos-London-resisting-multiple-rescue-attempts.html.

O'Brien, Keith, "Turkeys Take to Cities, Towns," *Boston Globe*, Oct. 23, 2007, www.boston.com/news/local/articles/2007/10/23/turkeys_take_to_cities_towns.

"Pachyderm Habitats," Denver Zoo: Special Exhibits, www.denverzoo.org/visitors/specialExhibits.asp.

"Penguin Picassos," ZooBorns, Feb. 1, 2010, www.zooborns.com/zooborns/2010/02/penguin-picassos.html.

"Pooch Plucked Twice from Ice," AP/WABC, March 17, 2010, http://abclocal.go.com/wabc/story?section=news/local/animals&id=7335017.

"Promotion for Japan's Stationmaster Cat," Japan Probe, Jan. 7, 2010, www.japanprobe.com/2010/01/07/promotion-for-japans-stationmaster-cat.

Pushkin, Yuri, "Ice Skating Bear Kills Russian Circus Hand," CNN, Oct. 23, 2009, www.cnn.com/2009/WORLD/europe/10/23/russia.skating.bear.death.

Robinson, Simon, "Monkey See, Monkey Do," *Time*, Oct. 17, 2006, www.time.com/time/world/article/0,8599,1546980,00.html.

Sahu, Sandeep, "Crowds Flock to Monkey 'Wedding,'" BBC News, Feb. 26, 2008, http://news.bbc.co.uk/2/hi/south_asia/7263782.stm.

Starmer-Smith, Charles, "Parrot That Tried to Mate with Mark Carwardine Is Given a Government Role," *Telegraph* (UK), Feb. 1, 2010, www.telegraph.co.uk/travel/travelnews/7128042/Parrot-that-tried-to-mate-with-Mark-Carwardine-is-given-a-government-role.html.

"Stationmaster Cat Draws Tourists," Japan Probe, April 23, 3008, www.japanprobe.com/2008/04/23/stationmaster-cat-draws-tourists.

"Taekwondo Monkeys Attack Trainer," *Telegraph* (UK), Dec. 16, 2009, www.telegraph.co.uk/news/newstopics/howaboutthat/6825449/Taekwondo-monkeys-attack-trainer.html.

"This Conductor's Got a Cat's Tongue," IOL News, May 25, 2008, www.iol.co.za/news/back-page/this-conductor-s-got-a-cat-s-tongue-1.401846.

"Turkeys Gone Wild: Aggressive Gobblers Attack Truckers at Michigan Auto Repair Shop," AP, March 21, 2009, www.chicagotribune.com/news/nationworld/sns-ap-odd-turkey-trouble,0,3373948.story (accessed Feb. 15, 2010).

"Turkey Terror in Rockport," *Gloucester (MA) Times*, Jan. 24, 2009, www.gloucestertimes.com/local/x645318182/Turkey-terror-in-Rockport-Post-Office-suspends-some-deliveries-after-birds-attacks-on-carriers.

Vanhoose, Joe, "Turkeys Terrorize Residents," *Athens (GA) Banner-Herald*, Jan. 27, 2010, www.onlineathens.com/stories/012710/new_555238477.shtml.

Vidal, John, "Giant Carnivorous Mice Threaten World's Greatest Seabird Colony," *Guardian* (UK), May 19, 2008, www.guardian.co.uk/environment/2008/may/19/wildlife.endangeredspecies.

Viegas, Jennifer, "Monkey Security Guards on Duty at the Commonwealth Games," *Discovery News*, Sept. 29, 2010, http://news.discovery.com/animals/monkey-security-guards-on-duty-at-the-commonwealth-games.html.

Vlahos, James, "Howl," *Outside Magazine*, Feb. 2009, http://outside.away.com/outside/culture/200902/monkey-problem-delhi-india-1.html.

Walker, Matt, "Movie Made by Chimpanzees to Be Broadcast on Television," BBC News, Jan. 25, 2010, http://news.bbc.co.uk/earth/hi/earth_news/newsid_8472000/8472831.stm.

Wellman, Walter, *The Aerial Age: A Thousand Miles by Airship Over the Atlantic Ocean*, Keller & Co., 1911.

"Wild Turkey Attack Traps Deputy in Cruiser," 2RSW, Nov. 16, 2009, www.nbc-2.com/global/story.asp?s=11514766.

"Wild Turkey Crashes through Window," UPI, March 18, 2010, www.upi.com/Odd_News/2010/03/18/Wild-turkey-crashes-through-window/UPI-81791268941407.

11. HEROIC HUMANS

"B.C. Logger Stuns Bear with Rock," CBC, Sept. 8, 2010, www.cbc.ca/canada/british-columbia/story/2010/09/08/bc-bear-rock.html.

Bedi, Rahul, "Indians Jail Marauding Monkeys," *Telegraph* (UK), Jan. 12, 2002, www.telegraph.co.uk/news/worldnews/asia/india/1381241/Indians-jail-marauding-monkeys.html.

"Billy the Kid Jailed in Germany," *Austrian Times*, Dec. 17, 2009, www.austriantimes.at/news/Around_the_World/2009-12-17/18983/Billy_the_kid_jailed_in_Germany.

Boyes, Roger, "Soon Every Swiss Dog Could Have His Day in Court," *London Times*, March 6, 2010, www.timesonline.co.uk/tol/news/world/europe/article7052004.ece.

"Cane Toads Meet Their Match in Tinned Cat Food," *Telegraph* (UK), Feb. 18, 2010, www.telegraph.co.uk/news/worldnews/australiaandthepacific/australia/7264897/Cane-toads-meet-their-match-in-tinned-cat-food.html.

Dys, Andrew, "Wild Goose Chase Nets Emu Near Downtown Rock Hill," *Rock Hill (SC) Herald*, May 5, 2010, www.heraldonline.com/2010/05/05/2143697/wild-goose-chase-nets-emu-just.html.

Flanagan, Jane, "Drunk Baboons Plague Cape Town's Exclusive Suburbs," *Telegraph* (UK), Aug. 29, 2010, www.telegraph.co.uk/news/worldnews/africaandindianocean/southafrica/7969313/Drunk-baboons-plague-Cape-Towns-exclusive-suburbs.html.

"Gardener Bites Snake to Death," *Borneo Bulletin*, Nov. 10, 2009, www.brusearch.com/news/53948.

Hadhazy, Adam, "Monkeys Go Bananas over Flying Squirrels," *Live Science*, July 30, 2010, www.livescience.com/animals/monkeys-attack-flying-squirrels-100730.html.

———, "Monkeys Hate Flying Squirrels, Report Monkey-Annoyance Experts," *Christian Science Monitor*, July 30, 2010, www.csmonitor.com/Science/2010/0730/Monkeys-hate-flying-squirrels-report-monkey-annoyance-experts.

Iannetta, Annamarie, "Montana Woman Fends Off Bear Attack with Zucchini," KECI/KCFW/KTVM, Sept. 23, 2010,http://www.nbcmontana.com/news/25130250/detail.html.

"Man Bites Snake in Epic Struggle," BBC News, April 15, 2009, http://news.bbc.co.uk/2/hi/7999909.stm.

"Meat Ants," Cane Toads in Oz, www.canetoadsinoz.com/meatants.html.

Mercer, Monica, "Car-Eating Dog Set Free," *Chattanooga Times Free Press*, March 26, 2010, www.timesfreepress.com/news/2010/mar/26/car-eating-dog-set-free.

"Monkey Urinates on Zambian President," AFP, June 24, 2009, www.google.com/hostednews/afp/article/ALeqM5iqYrQoDYoScCG_yLybzE_YuxTXBg.

"Montana Woman Fends Off Bear Attack with Zucchini," AP/*Billings (MT) Gazette*, Sept. 23, 2010, http://billingsgazette.com/news/state-and-regional/montana/article_ef1fedf2-c726-11df-8c9f-001cc4c03286.html.

"Pet Politics: Swiss to Vote on Whether Animals Need Lawyers," Spiegel Online International, March 3, 2010, www.spiegel.de/international/europe/0,1518,681363,00.html.

"Police Seize Parrot Trained as Drug Gang Look-out," ITN, Sept. 17, 2010, http://itn.co.uk/bdb1be466cc79c6c102b5cd15c140eda.html.

Schultz, Jason, "Suspect Arrested on Child Porn Charges Blames Cat," *Palm Beach Post*, Aug. 6, 2009, www.tcoasttalk.com/2009/08/06/suspect-arrested-on-child-porn-charges-blames-cat.

"Squirrel Banned from Riding Rollercoaster," *Telegraph* (UK), Feb. 15, 2010, www.telegraph.co.uk/news/newstopics/howaboutthat/7238666/Squirrel-banned-from-riding-rollercoaster.html.

Staight, Kerry, "Betrayal the Key to Feral Goats Fight," Australian Broadcasting Corporation, June 12, 2010, www.abc.net.au/news/stories/2010/06/11/2924956.htm.

"23 Injured by Monkeys in Shizuoka," *Japan Today*, Aug. 25, 2010, www.japantoday.com/category/national/view/23-injured-by-monkeys-in-shizuoka.

Willacy, Mark, "Japanese Police Hunt Rogue Monkey Gang," Australian Broadcasting Corporation, Sept. 2, 2010, www.abc.net.au/news/stories/2010/09/01/2999964.htm.

Williams, Alexandra, "Swiss Voters Reject Lawyers for Animals in Referendum," *Telegraph* (UK), March 8, 2010, www.telegraph.co.uk/news/worldnews/europe/switzerland/7394698/Swiss-voters-reject-lawyers-for-animals-in-referendum.html.

"Woman Survives Shark Attack by 'Punching, Punching, Punching,'" *Telegraph* (UK), Feb. 15, 2010, www.telegraph.co.uk/news/worldnews/australiaandthepacific/australia/7240211/Woman-survives-shark-attack-by-punching-punching-punching.html.

"Woman Tells Police Her Dog Made Her Steal Money," *Everett (WA) Herald*, June 30, 2009, www.heraldnet.com/article/20090630/NEWS01/706309908.

"Zambian President Evicts Monkeys from Residence After Urinating Incident," *Telegraph* (UK), Sept. 2, 2009, www.telegraph.co.uk/news/worldnews/africaandindianocean/zambia/6126010/Zambian-president-evicts-monkeys-from-residence-after-urinating-incident.html.

INDEX

abandoning young, ix, 43
Aberdeen, Scotland, 2
Abramson, Charles, 56
accidents and dogs, 108–9
Adélie penguins, 30–31, 50–51
Africa, 52, 67, 156
air travel and animals, 23–24, 147
AKC (American Kennel Club), 17
alarm calls, 76, 78–79, 80
albatrosses, 36, 146–47
alcohol abuse, vii, ix, 55–63, 64
Alfie (dog), 17
alligators, 16
American Kennel Club (AKC), 17
American West, 65
anacondas, 19
Anas platyrhynchos (mallard ducks), 36, 39
animal control officers, 132–34
animal lovers, 131
animals behaving badly, vii–xi
  assault, running amok, and arson, viii, 13–27
  devices and deceits, x, xi, 73–90
  family values, viii, ix, xi, 43–53
  heroic humans vs. bad animals, 155–69
  kinky creatures, viii–ix, xi, 29–42
  misdirection, vii, ix, 91–105
  muggers, burglars, and thieves, vii, viii, 1–12
  natural behavior, 123–36
  party animals, vii, ix–x, 55–70
  ungrateful beasts, 137–53
  See also dogs; *specific animals*
antelopes, 45–46, 78–79
ape vs. ape, devices and deceits, xi, 80–82
  See also specific primates
appearance, lying about, x, 88–89
aquatic risk factor for kinky behavior, xi, 37–38
Arizona, 69, 125, 130
Arkansas, 110
arsonists, viii, 16–17
art careers of animals, 150–51
Asia, 61
asking for it, natural behavior, 126–27
assault, running amok, and arson, viii, 13–27
Audubon Zoo in New Orleans, 86
Augsburg, Germany, 23
Australia, 14, 15, 18, 40, 44, 66, 69, 75, 86, 87, 91–92, 109, 134, 152, 165–67
Austria, 151

babirusas, 164
baboons
  heroic humans vs., 160–61
  kinky creatures, 32, 34
  muggers, burglars, and thieves, 3–4, 7, 8, 12

baboons (*cont.*)
natural behavior, 134
baby clothes and accessories eaten by dogs, 118–19
badgers, 61
badly behaving animals. *See* animals behaving badly
bands and pigeons, 26
bathroom, snakes in, 19, 21
BBC, 66, 150, 151
bears
heroic humans vs., 156–57, 168
muggers, burglars, and thieves, vii, 1, 2, 3, 4–5, 6–8
natural behavior, 128, 133
party animals, 60
ungrateful beasts, 144–45, 152
beauty as misdirection, 93–94, 155
bees, vii, ix, 55–57, 64, 118
beetles, 75
Belize, 63
Bensch, Victoria, 125–26
Berani (orangutan), 86
Berlin, Germany, 20
best man, bear as the, 152
billfish, 103
binge drinkers, 62
birds
assault, running amok, and arson, 24, 25–26
devices and deceits, 75
family values, 46, 49, 51
heroic humans vs., 11, 159–60
kinky creatures, ix, 30, 33, 34, 35, 36
misdirection, 93–94
party animals, 65–66
ungrateful beasts, 146–47
See also specific birds
biting by dogs, 108–9
blackbirds, 49, 65–66
black cockatoos, 162
black eagles, 47
Blackie (cat), 115–16
black rat snakes, 24
black-winged stilts, 33
blaming animals for human crime, 163
blue-footed booby chicks, 46
boars, 20–21
body-slamming by hummingbirds, 94
bonobos, 33, 98
Boo Boo (cat), 114
Boris (capybara), 142–43
Borneo, x, 88–89
Bourke, Brent, 41
brains of dogs, 119–21
Brazil, 11
breakfast cereal and chipmunks, 5, 6
breaking and entering, 3–5
breeding season and sex, 33
bridges to help dormice cross highway, 137
Britain, 5, 7
British Columbia, 93, 156
British MP toppled by cow, viii
broken wing deception by birds, 75
Bronx, New York, 19
Brookfield Zoo in Chicago, 131
Brookline, Massachusetts, 138–39
brown tree snakes, 145–46
bullying, 76
burglars, muggers, and thieves, vii, viii, 1–12
Busch (beer), 60
buzzards, 13–14

California, 15, 23, 51, 84, 92–93, 112
camels, 165
Cameron Park Zoo in Texas, 128
cane toads, 165–67
canines. *See* dogs
cannibalism and infanticide, viii, ix, 44, 45, 48, 95, 96, 97, 103, 165
Cape Town, South Africa, 3, 7
capuchin monkeys, 78, 83, 135–36
capybaras, 142–43, 165

cardboard door latch, 89
Caribbean, 60, 62
caribou, 33
cars and dogs, 113, 116–17, 121, 161
catnip, x, 63, 69
cats
assault, running amok, and arson, 16–17, 24, 25
devices and deceits, 77
dogs vs., 108, 114, 115
heroic humans vs., 163, 168
muggers, burglars, and thieves, 9–10
natural behavior, 132
party animals, 64
ungrateful beasts, 147, 148, 149
Centers for Disease Control and Prevention, 14, 108
chain saw bitten by a dog, 118
Chaos (dog), 113
Chapramari Wildlife Sanctuary, India, 21
Charbonneau, David, 130
Charlottesville, Virginia, viii, 9
Chattanooga, Tennessee, 161
Chernobyl disaster, 21
Chicago, Illinois, 131
chickens, 61, 143
Chihuahuas, 109
child abuse, 47–49
child porn blamed on cat, 163
child-saving dogs, 124–25
chimps
devices and deceits, xi, 75, 81, 83
dogs vs., 119
family values, 45
kinky creatures, 32–33, 34, 35, 40
misdirection, 96, 98, 99
party animals, 57
ungrateful beasts, 151
China, 30, 127, 141, 143–44, 152
chipmunks, 5, 6, 15
choosing your poison, 58–61
Christian Science Monitor, 168
Cincinnati Zoo, 128
circle of life, 123
circus animals, escaped, 22–23
climate change, 21
cocaine, 64
cockroaches, 66
Cocteau, Jean, 65–66
codependent creatures, 128–30
coffee, 67–68
cold weather and affection, 125–26
Colorado, 6–7, 150
Columbia, Brazil, 11, 159–60
Columbus, Ohio, 24
Commonwealth Games of 2010, 150
communication
devices and deceits, 76, 78–79, 80
dolphins, 100–101
communion received by dog, 152–53
concrete, breaking by chimp, 88
Connecticut, 49
Conniff, Richard, 93–94
conservation spokes-parrot, 150
coots, 48
corn snakes, 19
courtship by dolphins, 99–100, 104
cows, viii, 14–15, 16, 22, 55
crocs, 109
cross-species (interspecies) sex, 33–35, 40–42
crows, 15, 18, 24–25, 83, 85
cultural activities interference, 26
"cute" as misdirection, vii, 91–93, 95, 108, 126, 127, 131, 135, 155, 167, 169
cuttlefish, 74–75

dachshunds, 109
Daily Mail, 130–31
Dallas, Texas, 24
danger of nature, 123–24
Darwin, Charles, 34, 61–62
date rape, 37
Davies, Mr., 114

dead, mating with, 39–40
dead, playing, 75
Dead Duck Day, 39–40
death penalty, 160–61
deceits and devices, x, xi, 73–90
degus, 85
Delhi, India, 25
democratic process interference, 25
Denver, Colorado, 150
Department of Sustainability and the Environment, 134
devices and deceits, x, xi, 73–90
diamond thieves, dogs, 9, 111
Dino (bear), 138
dinosaurs, 96
Dipity (cat), 115
disabled veteran and service monkey, 135–36
disrespect, 10–12
dogs, viii, xi, 107–21
 accidents and dogs, 108–9
 assault, running amok, and arson, viii, 17, 24
 baby clothes and accessories eaten by, 118–19
 bees eaten by, 118
 biting, 108–9
 brains, 119–21
 cars and dogs, 113, 116–17, 121, 161
 cats vs., 108, 114, 115
 child-saving dogs, 124–25
 chimps vs., 119
 devices and deceits, 77, 82
 diamond thieves, 9, 111
 eating everything in sight, viii, 9, 111, 117–19, 163
 food tests with dogs, 119–20
 guns and dogs, viii, 110, 112, 121
 heroic humans vs., 140–41, 168
 injuries caused by dogs, 108–10
 money eaten by dogs, viii, 9, 111, 163
 party animals, 7, 59, 63–64, 65, 69–70
 police dogs, 113
 scavengers, dogs as, 120
 ungrateful beasts, 152–53
 "working" dogs, 112–13
Dogs Never Lie About Love (Masson), 78
dolphins, xi, 38, 97, 99–105, 168
door latch (cardboard), 89
Doritos, 2
drinking alcohol, vii, ix, 55–63, 64
drongos, 76
drug abuse, vii, ix–x, 55–70
drug-smuggling birds, 11, 159–60
Dublin, Ireland, 23
ducks, 36, 39
dwarf cavies, 32
dysfunctional families, ix
 See also family values

earwigs, 48
East Africa, 67
eating everything in sight, dogs, viii, 9, 111, 117–19, 163
"ecological citizenship," 134
educated animals, danger of, 143–45
egg-laying animals, 43
egrets, 47
Egypt, 143
electrical outages, 25, 145–46
electrified wires, avoiding, 86–87
elephants
 assault, running amok, and arson, 21–22, 22–23, 25
 family values, viii, 52–53
 heroic humans vs., 164
 muggers, burglars, and thieves, viii
 natural behavior, 128
 party animals, 58–60, 60–61, 65
elk, 32, 61
Ellie (dog), 118
employing animals, 147–53
emus, 158–59
enemies, our own worst, 123–36

England, viii, ix, 10, 12, 13, 15, 18, 23, 26–27, 50, 104, 112, 114, 128, 134, 157–58
Erie, Lake, 141
escaped circus animals, 22–23
experience, ignoring our, 130–32
"extra-pair paternity," 49
eyeglasses stolen, 9

Facebook, 151
family secrets, 96, 98–99
family values, viii, ix, xi, 43–53
fatherhood, ix
See also family values
favorites (family), 48
fearsome reputations as misdirection, 95–96
felines. *See* cats
"femme fatales," 79
fermentation as natural process, vii, ix, 55–56, 61
fiber-optic cable destruction, 24–25
fiddler crabs, 30
filmmakers, animals as, 151
firefighters rescuing dog, 140–41
fireflies, 79
fish, 74, 102, 103
"fish-aggregating devices," 102
flags stolen, 11–12
flamingos, 44
flat lizards, 74
flies, 66
floor squeegee escape, 89
Florida, 15, 19, 110, 117, 163
fly agaric mushrooms, 59
flying squirrels, 167–69
food, lying about, 76, 78, 79
food excuse, vii, 1, 2, 3, 7, 9, 18, 58
food tests with dogs, 119–20
football interference, 25–26
footwear fetish, viii, 9–10
foxes, viii, 9, 20
fox squirrels, 34
Frankfurt, Germany, 20
Fred (baboon), 8
friend, man's best. *See* dogs
frogs, ix, 22, 38, 43
fruit bats, 32, 63
fruit flies, 58
Fuji, Mount, 167
Fu Manchu (ape), 89–90
funnel web spiders, 37
fun vs. reproduction, 29, 35, 38, 39
fur seals, 34
future, planning for the, 87–88, 89–90

Gabon, Africa, 67
gambling, slot machine, viii
gangs, viii, xi, 52–53, 97, 100
gardener's revenge, 155–56
gay animals, 29, 34, 39–40, 50, 51, 152
geese, 34, 36, 41–42
Georges (dolphin), 104
Georgia, 139–40
Germany, 6, 9, 15, 20, 21, 22, 23, 61, 84, 159
gibbons, 128
Gibson, Jerry, 158
gloves stolen, 9–10
goats
assault, running amok, and arson, 15, 16
heroic humans vs., 159, 162–63, 165
party animals, 55, 65, 67–68
golden lion tamarins, 33
Goodall, Jane, 96
gorillas, 67
Gough Island, South Pacific, 146
government action on animals behaving badly, 161–62
grand theft animal, 6–8
graylag geese, 34
gray squirrels, 34, 36, 134
great tits, 76
Greece, 22
Gregson, Paul, 17

groundhogs, viii
ground squirrels, 80
grouse, 34
Guam, 145–46
guanacos, 164
guarding livestock, 112
guns and dogs, viii, 110, 112, 121
Guwahati, India, 61

Hambone Award, 117–18
Hammond, David, 142–43
Hamric, Bill, 136
Hamric, Joseph, 135–36
hamsters, 44
hangovers, 61–63
Hanuman, 149–50
harbor porpoises, 101, 103
Hawaii, 25–26, 59
Hawaiian monk seals, 39
Hawaii News Now, 26
Heavy Lamar (orangutan), 90
"helpful" males, 52
helping distressed creatures, 132–34
Henry, Paul, 16
heroic humans vs. bad animals, 155–69
herring gulls, 44, 48
"hippie monkey" (muriqui), 98–99
Hogben, Ann, 116
Ho Hos, x, 70
home invasions, 3–5
home turf, 18–19
homosexuality, 29, 34, 39–40, 50, 51, 152
honeybees, vii, ix, 55–57, 64, 118
Honeymoon Ridge, Australia, 40
Honolulu, Hawaii, 142
horses, 31, 65
house mice, 146–47
house sparrows, 45
Houston, Texas, 24
howler monkeys, 164
Huffman, Professor, 67
human garments and accessories stolen, viii, 9–10
humans (heroic) vs. bad animals, 155–69
human vices and crimes, animals doing, viii–x
hummingbirds, viii, 30, 33, 36, 93–94
hunger excuse, vii, 1, 2, 3, 7, 9, 18, 58
hyenas, 46

ice-skating bear, 144–45
Idaho, 117
Ig Nobel Prize for biology, 39
Illinois, 41, 131, 139
impairments from substance abuse, 56, 57, 58, 59, 60, 61, 62, 63, 64, 65, 66, 67, 68, 69
India, 9, 21–22, 23, 25, 60–61, 66, 152, 155–56, 160
individuals vs. bad animal behavior, 155–57
infanticide and cannibalism, viii, ix, 44, 45, 48, 95, 96, 97, 103, 165
infrastructure disruptions, 21–25, 145–46
injured, playing, 75
injuries caused by dogs, 108–10
"innocent" wild things, vii
intelligence hypothesis, Machiavellian, 78
Internet access interference, 24–25
interspecies (cross-species) sex, 33–35, 40–42
invasive species, 91–92, 134, 145, 162
Iowa, 16
Iran, 11
Ireland, 23
Italian sandwich mugging in New Jersey, vii, 2
Italy, 138

jackdaws, 26–27
Jack Russell terriers, 109
jaguars, 128
Japan, 24–25, 33, 148–49, 167–68
Jean-Felix and wild turkeys, 138

jellyfish, 109
Jimmie (chimp), 81
"Judas goats," 162–63, 165
junk food junkies, x, 70–71

kakapos, 150
kangaroos, 14, 18, 31, 32, 40–41, 44, 152
Karta (orangutan), 87
Katherine, Australia, 69
Kauai, Hawaii, 25–26
kea parrots, 10
Kentucky (penguin), 128–29
Kenya, 156
kestrels, 35
Khat (qat), 67
Kiddo (cat), 147
Kings of Leon, 26
kinky creatures, viii–ix, xi, 29–42
kittiwakes, 46
Korea, 24
Kyrgyzstan, 144–45

Laing, Joan, 3
Lana (cat), 115
langur monkeys, 45, 149–50
law enforcement for animals behaving badly, 159–61
lawn ornaments and sex, 41–42
laws and ungrateful beasts, 148–49
laws protecting animals, 138–40
laziness of lions, 95
leaf edges used as tools, 85
learning family values, 45–47
leaves to make voices deeper, x, 88–89
lemurs, 68
lesson learned, natural behavior, 135–36
lightning, 109
Lincoln Park Zoo, 98
lions, viii, 35, 44, 95, 104, 128
livestock guarding, 112
lizards, 66, 74
llamas, 23
locoweed, 65
Long Beach, California, 92–93
Long Island, New York, 116
Lorenzo (parrot), 159–60
Louisiana, 86, 87
love, lying about, 78–79
Luang, Hu, 144
Lulu (dog), 118–19
luring humans, xi, 81
lying, x, 73, 74–75

macaques
  devices and deceits, 82
  heroic humans vs., 167–68
  kinky creatures, 30, 32–33, 34
  natural behavior, 130–32
  ungrateful beasts, 149–50
Machiavellian intelligence hypothesis, 78
Magic (cat), 115
magnificence as misdirection, 95–96
magpies, 6
mahi-mahi, 102
mail delivery interference, 24, 114–16, 139
Maine, 22
Malaysia, 63
mallard ducks *(Anas platyrhynchos)*, 36, 39
manatees, 38
Mangrum, Bobby, 159
man's best friend. *See* dogs
marine iguanas, 31
marmosets, 32, 164
Martin, Joanne, 41
Maryland, 9, 111
Massachusetts, 138–39
Masson, Jeffrey Moussaieff, 78
masturbation, ix, 29, 31–32
maternal instinct, ix
  See also family values
MEarth project, 130
meat ants, 166–67
meerkats, 45, 75, 150–51
Melbourne, Australia, 15

Memorial Day disruption, 11–12
Memphis, Tennessee, 25
menopause and sex, 33
meow-purr, 77
Mexico, 59
mice, 66, 146–47
Michigan, 11–12, 139
Middle East, 67
Minnesota, 117, 131
Minnesota Zoo, 131
misdirection, vii, ix, 91–105
Missouri, 26
money eaten by dogs, viii, 9, 111, 163
monkeys
  assault, running amok, and arson, 25
  devices and deceits, 78, 80, 82, 83
  family values, 45, 47–48
  heroic humans vs., 160, 161, 164, 167–68
  kinky creatures, 30, 32, 33
  misdirection, 98–99
  muggers, burglars, and thieves, 9, 11
  natural behavior, 130–32, 135–36
  party animals, 55, 60, 62, 66
  ungrateful beasts, 143–44, 148, 149–50, 152
  See also specific monkeys
monogamy, 49, 51–52
Montana, 3, 152, 156–57
moose, 22, 33
morphine, 64
mountain sheep, 38
mounting chicks by parents, 48
muggers, burglars, and thieves, vii, viii, 1–12
multitalented marauders, 20–21
muriqui ("hippie monkey"), 98–99
mushrooms, 59
mute swans, 34

National Geographic, 151
natural behavior, 123–36
naturalists, 134
Natuurhistorisch Museum Rotterdam, 39
Nazlat Imara, Egypt, 143
Nebraska, 89
necrophilia, 39–40
neglect, 44
New Caledonian crows, 83, 85
New Delhi, 149
Newell's Shearwater birds, 25–26
New Hampshire, 4
New Jersey, vii, 2, 129
newspaper reading interference, 26–27
New York, 19, 116
New York City mayor bitten by groundhog on Groundhog Day, viii
New York Times, xi, 56, 57, 99
New Zealand, 10, 110, 113, 150
Noah (service monkey), 135–36
"non-human persons" (dolphins), 97
North Beaver Township Fire Department, 16
North Carolina, 9, 111
Northern Territory News, 70
Nottinghamshire, England, 26–27
Nshi, China, 144

"obligate siblicide," 46–47
octopus, 84
Of Mice and Men (Steinbeck), 111
Ohio, 24, 128
Ohio State University, 56
Oklahoma, 22, 113
olive baboons, 32
Olson, Ms., 92–93
Omaha Zoo, 89
omission, sins of, 80
opium, 64, 65–66
oral sex, 33–34
orange hiding, 80–81
orangutans
  devices and deceits, x, 74, 80–81, 81–82, 85, 86–87, 88–89

kinky creatures, 32–33, 34
ungrateful beasts, 151
Oregon, 110, 150
Orlando, Joe, 129
Oscar (cat), 10
ostrich, 158
otters, 19, 24, 126, 128
Otto (octopus), 84
our own worst enemies, 123–36
oystercatchers, 35, 44

pachyderms. *See* elephants
paintings by animals, 150–51
pandas, 30, 126–27, 168
parenting, 43, 47–49
See also family values
Parnell, Peter, 50
parrots, 10, 150, 159–60
partners in crime, 11
party animals, vii, ix–x, 55–70
passport stolen, 10
Patagonia, 164
penguins
family values, 50–51
kinky creatures, 30–31, 34–35, 37, 39
natural behavior, 128–29
ungrateful beasts, 150–51, 152
penis, snake biting a man's, 19
Pennsylvania, 139
perversion, viii–ix, xi, 29–42
"Peter Pan" orangutans, 74
pets and fooling humans, 77
peyote, 65
pheromones, 37
Philadelphia, Pennsylvania, vii
physical deception, 73–75
Pickering, Megan, 69–70
pigeons, 11, 26, 64
pigs, 46, 55, 61, 141–42, 164
pigtail macaques, 48
pit bulls, 109
planes and animals, 23–24, 147
planning for the future, 87–88, 89–90
platypus, 109
Poland, 19
police, monkey serving on, 149
police dogs, 113
political process interference, 25
politicians on animals behaving badly, 161–62
popularity of alcoholic beverages, 58
porcine diamond thieves, 9
porcupines, 33
pornography, 30
porpoises, 101, 103
Portswood, Southampton, England, 10
possum, playing, 75
pot, 63–64
prairie voles, 51–52
pregnancy and sex, 33
premeditation, 7
proboscis monkeys, 33
promiscuity, 49, 50, 51–52
pronghorn antelopes, 45–46
prostitution, 30–31
Pryor, Karen, 99, 103–4
psilocybin mushrooms, 59
psych out, 89–90
Punjab, India, 160
pushers (drug), 67–68
pythons, 156

qat (Khat), 67
Queensland, Australia, 18
quinine, 58

rabbits, viii, 91–93, 145
raccoons, 25
Radford University in Virginia, 37
Rainier (beer), 60
rakes used as tools, 85
Ramesh (gardener), 155–56
Ramsgate, Kent, 116

rape, viii, xi, 37–38, 45, 94, 97, 100–101
rats, x, 16, 23, 70, 71, 77
rays, 102
realty TV genre, 13
Red (dog), 7
red-tail monkeys, 33
reindeer, 59
reproduction vs. fun, 29, 35, 38, 39
rescue missions, 140–43
Reuters, 57
rhesus monkeys, 47–48, 80
rhinos, 52, 150
rice beer, 60–61, 61
Richardson, Justin, 50
right whales, 38
ring-billed gulls, 48
ringtail cats, 130
Rock Hill, South Carolina, 158–59
Rockport, Massachusetts, 139
rocks used as tools, 83
roller coaster, 157–58
rooftop luggage boxes stolen, 12
roosters, ix, 79
Roscoe (dog), 118
Rouwendal, Henry, vii, 2
running amok, assault, and arson, viii, 13–27
running wild (unexpected animals), 22–23
Russia, 57, 59

Sacramento, California, 23
salt marsh sparrows, 49
sand tiger sharks, 46
San Francisco, California, 51
San Jose, California, 15
San Pedro, California, 84
Santa Monica Aquarium, 84
Santini (chimp), 87–88
Scarborough, South Africa, 160–61
scavengers, dogs as, 120
scientific solutions to animals behaving badly, 162–68
Scotland, 101, 142, 151
screwing around (promiscuity), 49, 50, 51–52
scrub fowl, 43
seabirds, 36, 146–47
seagulls, 1, 2
sea lions, 44, 150
seals, 34–35, 37–38, 39, 44
Seattle, Washington State, 9, 63–64
sea turtles, 102, 103
security force, monkey serving on, 149–50
self-sacrifice vs. self-interest, 124–26
semiaquatic risk factor for kinky behavior, xi, 37–38
service monkey, 135–36
sex crimes, viii, ix, xi, 35–40, 45, 94, 97, 100–101, 150
sex objects, 32–33
sharks, 102, 109, 132, 157
sheep, 38–39, 66
"she-males," 74
Shine, Rick, 166
shoes stolen, 9
Siberia, 59
sibling killers, 45–47
Siegel, Ronald, 58–59, 60, 65
Simon's Town, South Africa, 8
Singer, Fred, 37
single teen mothers, viii, 52–53
sins of omission, 80
Sky News, 11
slot machine gambling, viii
Smithsonian Magazine, 95
Smithsonian's observatory in Arizona, 130
smoking habits, 57
snakes
  assault, running amok, and arson, 19, 24, 25
  devices and deceits, 73–74
  dogs vs., 109
  heroic humans vs., 155–56
  ungrateful beasts, 145–46

sneaker males, 74, 75
sneaking, 82
snow leopards, 128
social climbing, 45
social drinkers, 62
society vs. bad animal behavior, 157–59
Somerset, England, 50
Sonic Spinball roller coaster, 157–58
South Africa, 7, 8, 76, 134, 160–61
South America, 164
South Australia, 162
South Carolina, 111, 158–59
sparrows, 45, 49
spider monkeys, 32
spiders, 66, 75, 109
spies, pigeon, 11
squirrels
  assault, running amok, and arson, 15, 23–24, 25
  devices and deceits, 80
  heroic humans vs., 157–58, 168
  kinky creatures, 34, 36
  muggers, burglars, and thieves, 11–12
  natural behavior, 134
  ungrateful beasts, 142
St. Georges, Joe, 141
St. Kitts, 60, 62
St. Louis, Missouri, 26
Staffordshire, England, 158
stationmasters, animals as, 148, 149
steady drinkers, 62
Steinbeck, John, 111
Stevenson, Adam, 128–29
sticks used as tools, 83
stones, throwing at zoo tourists, 87–88
stove knobs, covering, 17
substance abuse, vii, ix–x, 55–70
sugarcane, viii, 56
Sun (newspaper), 5, 6
swallows, 34, 35, 39, 78
swans, 49, 50, 140
sweater stolen, viii, 9
Sweden, 61, 87–88
Switzerland, 22–23, 161–62

T. rex *(Tyrannosaurus rex)*, 96
Tae kwon do, 144
Taiwan, 19
Tama (cat), 148, 149
Tammy (wild turkey), 129
Tango Makes Three, And (Richardson and Parnell), 50
Tanzania, 96
technological advances, 32–33, 82–89
teenage mothers, viii, 52–53
teetotalers, 62
Tehran, Iran, 11
temper tantrums, 47–48
Tennessee, 16, 25, 161
Texas, 11, 24, 47, 128
Thailand, viii, 130, 149
thieves, muggers, and burglars, vii, viii, 1–12
"Things that turned out to be other things," 133–34
Thorndike, Edward, 119
three-strikes policy, 160–61
Tiger (cat), 115
tigers, 165
tiny terrors, viii, 5–6, 15
Tirath, Ram, 160
toad-licking, 68–70
toilets, snakes in, 19, 21
Tokioka, Jimmy, 26
Tokyo, Japan, 24–25, 148
tools, using, 32–33, 82–89
topi antelopes, 78–79
Toronto, 152
toxoplasma, 77
tradition, treading on, 25–27
traffic problems, 21–24, 61, 113, 129, 139, 140, 159
trains, blocking, 23
traveling animals, 145–47

tree shrews, 63
tree swallows, 35
triggerfish, 103
tuna, 102
turkeys, vii, 129, 138–40
Tyrannosaurus rex (T. rex), 96

umbrellas and lions, 95
underwear stolen, 10
unexpected (running wild) animals, 22–23
ungrateful beasts, 137–53
uninvited company, 3–5
United Nations, 132
University of Cape Town's Baboon
Research Unit, 134
University of Sydney, Australia, 166
University of Victoria, British Columbia, 93
U.S. Geological Survey (USGS), 145–46
U.S. Senate, 132
using tools, 32–33, 82–89
Uttar Pradesh, India, 66

valuables eaten by dogs, viii, 9, 111, 163
vehicle theft, 6–8
Vernon, New Jersey, vii, 2
vervet monkeys, 60, 62, 80
Vienna, Austria, 151
Vietnam War, 135
vineyard owners and baboons, 134
violence, 96, 98–99, 101, 103
Virginia, viii, 9, 139
voices deeper, using leaves for, x, 88–89
voles, 51–52

Waal, Frans de, xi, 81
wahoo, 102
waiters, monkeys as, 148
Wales, 17, 137
wallabies, 23, 66
walruses, 31
Washington Post, 131–33
Washington State, 9, 16–17, 60, 63–64, 139
water disliked by penguin, 128–29
water moccasins, 19
weddings, animals participating in, 152
Welcome Glen Baboon-Free
Neighbourhood Action Group, 3
West Bengal, India, 21–22
West Virginia, 23–24
whales, 38
White, Thomas, 97
Who's Who in the Zoo: A Natural History of Mammals (WPA Federal Writers' Project), 164–65
wiggle dance of honeybees, 64
wild boars, 20–21
wild turkeys, vii, 129, 138–40
wild (unexpected) animals, 22–23
William (baboon), 160–61
wine industry and baboons, 134
Winston (dog), 161
wire used as tools, 83, 85, 89–90
Wisconsin, 128
Witt, Illinois, 41
wolves, 120, 131
wombats, 134
"working" dogs, 112–13
WPA Federal Writers' Project, 164–65
wrong place, wrong time, 19, 21–22
Wung, Lo, 144

Yang Yang (panda), 127
yellowtail, 102
Yogi (cat), 114
Yorkshire, Maryland, 9

Zambia, 161
zebras, 23
Zhora (chimp), 57
zoo patrons, 126–28, 131
zucchini weapon used against bear, 156–57
Zurich, Switzerland, 22–23